Patrick Moore's
Practical Astronomy Series

Other titles in this series

Using the Meade ETX

100 Objects You Can Really See with the Mighty ETX

Mike Weasner

With contributions by P. Clay Sherrod
and Richard (Dick) Seymour

With 74 Figures

Springer

Cover illustrations: ETX-90EC courtesy of Meade Instruments Corporation. Background image by Mike Weasner using a 35 mm film camera mounted piggyback on an ETX-90RA. Moon by Mike Weasner using a Casio digital camera and an ETX-90RA. Insets: M42 photograph by Gerald Wechselberger, Ebreichsdorf, Austria, using an ETX-125EC and Olympus Z2020 digital camera. Used with permission. Jupiter and Saturn photographs by Rick Krejci, Scottsdale, Arizona, USA, using an ETX-90 and Logitech QuickCam VC webcam. Used with permission.

British Library Cataloguing in Publication Data
Weasner, Mike
   Using the Meade ETX: 100 objects you can really see with the
   mighty ETX. – (Patrick Moore's practical astronomy series)
   1. Telescopes 2. Astronomy
   I. Title
   522.2
ISBN 1852333510

Library of Congress Cataloging-in-Publication Data
Weasner, Mike, 1948–
   Using the Meade ETX: 100 objects you can really see with the mighty
   ETX / Mike Weasner.
     p. cm. – (Patrick Moore's practical astronomy
     series, ISSN 1617-7185)
     ISBN 1-85233-351-0 (acid-free paper)
     1. Reflecting telescopes. 2. Catadioptric systems. I. Title. II. Series.
QB88. W38 2001
522′. 2–dc21   2001045963

Patrick Moore's Practical Astronomy Series ISSN 1617-7185
ISBN-10: 1-85233-351-0 Printed on acid-free paper
ISBN-13: 978-1-85233-351-5

© Springer-Verlag London Limited 2002

9 8 7 6 5

Springer Science+Business Media
springer.com

Dedication

This book might have been written by anyone, but the fact that I wrote it is a direct result of my older brother Paul. When I was six years old, he began teaching me the night sky and so started my life-long interest in astronomy. Later he got me interested in science fiction and much later, computers. Without his interest in these things and his influence on me, my life would have been much different. I followed Paul (by several years) to Indiana University, where I obtained my B.S. in Astrophysics. Paul passed away unexpectedly in June 2000 as I was working on this book. He was thrilled that I was writing a book on one of his favorite subjects. His support, knowledge, and love were major influences in my life. He is sorely missed. Thanks Paul, for everything.

Paul Weasner
15 June 2000

Preface

Meade Instruments Corporation introduced the original ETX Astro Telescope in the Spring of 1996. I purchased one in September 1996 and soon realized just how mighty this little telescope was. A few days later, I created the first ETX-only Site on the World Wide Web (http://www.weasner.com/etx). I am proud to say that this Site has since become the Internet's most popular ETX Site. Even though the ETX performs like a much larger telescope (as implied by the logo from my Web Site, Figure P.1), it really is just a 90 mm (3.5-inch) telescope.

In 1999, Meade released two new models: the ETX-90EC in January and the ETX-125EC in May. Both of these new models kept the original optical design of the ETX but added a new, exciting, computerized capability to make finding objects in the sky much easier than before. Upon the release of the EC models, the original ETX became known as the ETX-90RA. In May 2000, Meade announced the ETX-60AT and ETX-70AT models. Unlike the earlier ETX models, which are

Figure P.1. Mighty ETX Web Site graphic.

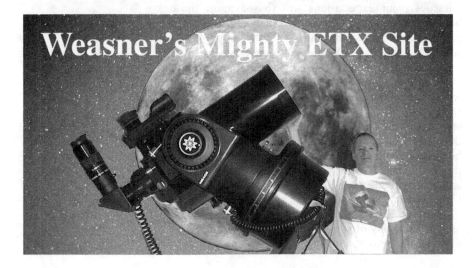

Maksutov–Cassegrain telescopes, these models are refractor telescopes (no mirrors) but have the same low-cost computerized object-finding capability.

There are a lot of books on basic astronomy and telescope usage for the casual and semiserious amateur astronomer, but this book is dedicated to the ETX suite. With the original ETX, Meade created a new market with a relatively inexpensive, small yet capable telescope. Its optical design provided a long focal length telescope (allowing for higher magnifications) in a tube of very short length. Due to its small size, it was portable, meaning you could set it up on a moment's notice or take it on trips. It had a motorized right ascension drive for tracking the movement of the sky due to the Earth's rotation. It had excellent optics for its size and cost, thereby providing views that kept you coming back for more. Meade built upon the popularity of the original model with the EC models. They added motorized control on both the right ascension (or azimuth) axis and the declination (or altitude) axis. They also added an optional handheld computer, originally containing some 12 000 objects in its database. This computerized handcontroller allows you to select an object from the database (or enter by hand) and the ETX will slew (GOTO) to place it in the eyepiece. No more being lost in space trying to locate faint or unfamiliar objects.

In the ads for the ETX models and all the books on astronomy and telescopes, you see beautiful astrophotographs of the Moon, the planets, galaxies, and gaseous clouds (called nebulae). When you buy any telescope, your expectations will (based upon those beautiful photographic images) usually exceed what you will actually see when using the telescope. While some objects, like the Moon, will look very similar to the photographs, most celestial objects will not. As you move to larger telescopes, typically the more details you will see and the fainter the objects that can be seen. But reality usually sets in fairly quickly. Your expectations were too high. This book should assist you in setting those expectations for the ETX. I will discuss 100 of the best visual objects, some with drawings or photographs that attempt to illustrate what your eye will actually see, and I will provide some hints, tips, and techniques that will develop your eye to best see these objects.

After you have experienced the enjoyment of visual astronomy, many of you will want to take photographs of what you see. You can take excellent photographs

Figure P.2.
www.weasner.com/etx

using the ETX with almost any camera, and this book will discuss doing just that.

There is a whole "universe" of ETX users "out there" and there are many places to get information and help. I will discuss some of the resources that are available to the ETX user to get the most out of the telescope.

This book is essentially an expansion of a chapter on the ETX that appears in *Astronomy with Small Telescopes*, edited by Stephen F. Tonkin, and published in 2001 by Springer in its Practical Astronomy Series. If you want to learn about several small telescopes and see how the ETX compares, check out that book. If you are considering an ETX or already have one, the book you have in your hands will show you why I call it "The Mighty ETX" and demonstrate how you can get a lot of enjoyment and education from the ETX.

Disclaimer: Companies, products, and Web Sites mentioned in this book may no longer exist or be available by the time you read about them. Their mention here should not be taken as an endorsement.

Comments, suggestions, and corrections are welcome. They may be included in a future edition of this book. I will post them on my ETX Web Site, if appropriate. The best way to contact me is via electronic mail. If you do not have email, you can send feedback to the publisher, who will forward it to me.

Thanks. Enjoy this book. And enjoy your telescope, whether it is an ETX or some other small telescope. The Universe is waiting for you.

Michael Weasner
mweasner@aol.com
Rolling Hills Estates, California, USA
November 2001

Acknowledgements

Books take a lot of work (I now know). Without the efforts of a lot of people, you would not be reading this one. The idea for this book originated with John Watson, Astronomy Editor and CEO of Springer-Verlag London and an ETX owner. I hope this book at least comes close to matching the concept he had when he asked me to write the book. My wonderful and intelligent wife, R-Laurraine Tutihasi, did the illustrations (unless otherwise credited), drawings at the eyepiece, and proof-reading. Without her support and abilities, I would not have even attempted to write this book. Dr. Leon Palmer, Rigel Systems, reviewed the draft manuscript and made many valuable comments. The improvements are due to him; any areas that could still be improved are totally mine. As this book was going to press, it was decided to add some specific and valuable content from my ETX Web Site. Clay Sherrod, the Site's optical and mechanical expert, and Richard Seymour, the Site's Autostar expert, agreed to the inclusion of their material in Chapter 6. Thanks guys. I would like to thank Meade Instruments for developing the ETX and for their continued support of Weasner's Mighty ETX Site. Lastly, I would like to thank all the visitors and contributors of information, tips, and photographs to my ETX Site. Their enthusiastic participation makes running the Site fun and worthwhile. Some of the tips and information in this book come from the Site visitors. This book is their reward.

Contents

Chapter 1

ETX Basics

Prior to the ETX, most inexpensive consumer telescopes came in two varieties: the refractor telescope, which uses a glass lens at the top to focus the light, and the Newtonian reflector telescope, which uses a curved mirror (actually a spherical surface) at the bottom to reflect and focus the light. Other telescopes were available that used alternative designs, but they were typically expensive and not appropriate for the mass-market amateur astronomer. The Maksutov–Cassegrain design of the original ETX was used previously in the highly sought after but pricey Questar 3.5-inch telescope. The original ETX brought this efficient design to many more amateur astronomers at an affordable price. The Maksutov–Cassegrain design uses a lens and mirrors to focus the light, as shown in Figure 1.1. Light enters the telescope through a coated corrector lens at one end of the ETX optical tube assembly (OTA). It then proceeds down the tube to a mirrored surface (the primary mirror), where it is reflected back up the OTA only to be reflected again from a small "spot" (secondary) mirror on the inside of the corrector lens. This last reflection sends the light back towards the rear of the ETX OTA where, using a "flip mirror", it is directed to the normal eyepiece position or out the rear port on the back of the ETX tube. This multireflected light path results in the compact design of the ETX, yielding a much longer focal length (optical length) than is otherwise possible. The use of high-quality optics yields the impressive views that you see through the ETX.

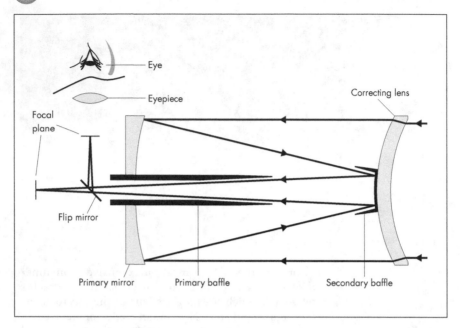

When you view objects through an eyepiece in the normal position on the top of the ETX tube, they appear right side up but reversed left-to-right. Figure 1.2 shows how the word "Meade" would appear when viewed through an eyepiece. When looking at terrestrial objects, this image reversal can be undesirable. You can attach an optional "erecting prism" to the rear port on the ETX and insert an eyepiece into it. Objects appear upright and not reversed when viewed through the erecting prism.

The mount itself has two fork arms that support the telescope tube (just as with the Hale 200-inch Telescope on Mount Palomar). The fork arms are fixed to a rotating base that contains the electronics and motorized drives that control the two axes. The mount is made of plastic and metal, allowing it to be both strong and lightweight. This makes the ETX a true portable telescope.

Figure 1.1.
Simplified drawing of the Maksutov–Cassegrain design.

Figure 1.2. Reversed view through eyepiece.

Specifications

The following table shows the specifications for the ETX models discussed in this book. The information comes from the Meade ETX manuals.

	ETX-90RA	ETX-90EC	ETX-125EC
Optical design	Maksutov–Cassegrain	Maksutov–Cassegrain	Maksutov–Cassegrain
Clear aperture	90 mm (3.5-inch)	90 mm (3.5-inch)	127 mm (5-inch)
Focal length	1250 mm (49.2-inch)	1250 mm (49.2-inch)	1900 mm (74.8-inch)
Focal ratio (photographic speed)	f/13.8	f/13.8	f/15
Near focus (approx)	3.5 m (11.5-feet)	3.5 m (11.5-feet)	4.6 m (15-feet)
Resolving power	1.3 arc secs	1.3 arc secs	0.9 arc secs
Super multicoatings	standard	standard	standard
Limiting visual stellar magnitude	11.7	11.7	12.5
Maximum practical visual power	325×	325×	500×
Optical tube dimensions (diameter × length)	10.4 cm × 27.9 cm (4.1 inch × 11 inch)	10.4 cm × 27.9 cm (4.1 inch × 11 inch)	14.6 cm × 36 cm (5.75 inch × 14.2 inch)
Secondary mirror obstruction	27.9 mm (1.1-inch); 9.6%	27.9 mm (1.1-inch); 9.6%	39.4 mm (1.6-inch); 9.6%
Telescope mounting	fork type; double tine	fork type; double tine	fork type; double tine
RA drive motor	4.5 volts DC	12 volts DC	12 volts DC
hemispheres of operation	north and south; switchable	north and south; switchable	north and south; switchable
Slow-motion controls	manual; RA and DEC	electric; 4 speed; RA and DEC	electric; 4 speed; RA and DEC
Bearings	RA and DEC: Nylon	UHMW polyethylene (DEC); PTFE (RA)	UHMW polyethylene (DEC); PTFE (RA)
Materials:			
Tube body	aluminum	aluminum	aluminum
Mounting	reinforced high-impact ABS	reinforced high-impact ABS	zinc reinforced high-impact ABS
Primary mirror	Pyrex glass	Pyrex glass	Pyrex glass
Correcting lens	BK7 optical glass, Grade-A	BK7 optical glass, Grade-A	BK7 optical glass, Grade-A
Telescope dimensions	38 cm × 18 cm × 22 cm (15 inch × 7 inch × 9 inch)	38 cm × 18 cm × 22 cm (15 inch × 7 inch × 9 inch)	48 cm × 23 cm × 27 cm (19 inch × 8.9 inch × 10.8 inch)
Telescope net weight (90RA incl. Tripod legs; 90EC incl. electronic controller and batteries)	4.2 kg (9.2 lbs)	3.5 kg (7.8 lbs)	8.5 kg (18.8 lbs)
Telescope shipping weight	5.6 kg (12.4 lbs)	5.6 kg (12.4 lbs)	10.8 kg (23.7 lbs)
Price (US$)	$595	$595	$895

ETX-90RA

The original ETX was released in 1996. It was the first widely available consumer telescope that combined excellent optics, small size, and a battery-operated motor-driven right ascension drive to compensate for the Earth's rotation, all at a reasonable and affordable price. Many users even loved the way it looked, with its compact design, deep purple tube, and flat black fork mount and base. Here was a quality telescope that many could afford and use. It came standard with the motorized right ascension tracking drive, a 26 mm 1.25-inch eyepiece (yielding a 48× magnification), 8 × 21

Figure 1.3. Original ETX. Photo courtesy of Meade Instruments Corp.

mm finderscope, tabletop tripod legs that were adjustable for the user's latitude, and right ascension and declination setting circles. By using the standard 1.25-inch eyepiece size, there was almost no limit to the accessories that were immediately available for use with the ETX. These included other focal length eyepieces, Barlow lenses, filters, and various types of adapters. Both the base and the OTA had standard $\frac{1}{4}$-inch tripod mounting attachment holes. This allows the entire ETX (minus the legs) or just the OTA to be mounted on a sturdy camera tripod. The ETX included a rear port where the user could attach other accessories, such as erecting prisms, camera adapters (for prime focus or eyepiece projection photography), eyepiece holders, and wide-field adapters. Thanks to the popularity of the ETX and in part to address some of its shortcomings, several companies developed modifications or add-ons specifically for the ETX. I will discuss many of these accessories and add-ons later in this chapter.

The original ETX was also available in a lower-priced "spotting scope" version. This scope was intended for terrestrial use only and so did not include the fork mount and motorized base but did include an erecting prism. Several later models of the ETX line are also available in spotting scope versions that include just the OTA, a low-power eyepiece, and the erecting prism.

When first released, the ETX had an introductory suggested retail price of $495 (all prices are given in US dollars) but within a few months this was raised to its current suggest retail price of $595.

When the next ETX model, the ETX-90EC, was released, the original ETX had a name change from just "ETX" to "ETX-90RA". The "-90" denoted its aperture (90 mm) and the "RA" meant it had a right ascension drive. In some markets, this model was also known as the "ETX Astro Model M".

ETX-90EC

In January 1999, the ETX world changed. Meade announced an improved ETX: the "ETX-90EC". The "EC" stood for "electronic control", and indeed this new ETX model has just that. The basic telescope design (telescope tube, optics, and fork mount) was the same as the original ETX, but this new model has the advantage of having motor-driven tracking and slewing in both right ascension and declination. No longer was it necessary to manually move the telescope to point to an object; you used the included handcontroller to move (known as "slew") the telescope at one of four speeds (from slow to high speed). Once you stopped slewing, the RA drive would automatically begin tracking. But Meade did not stop there.

Since the ETX could now be slewed electronically in both RA and DEC (declination) or azimuth and altitude, Meade also announced in January 1999 the "Autostar" computerized "GOTO" controller (model #497). This optional accessory for the ETX-90EC replaced the standard handcontroller. The Autostar's computer can be used to automatically point the telescope to any object in the Autostar's database of 14 000 objects. With the Autostar, the ETX-90EC can be set upon any flat, level surface in an altitude/azimuth orientation, no longer requiring that the telescope be mounted in a polar (also known as "equatorial") aligned mode. The computer will drive the telescope in both axes to accurately track objects, even Earth-orbiting satellites. Since this alt/az tracking capability now existed, the three tabletop tripod legs became optional. The Autostar software itself can be updated using a personal computer and an optional serial cable. You can slew the telescope with the Autostar handcontroller at one of nine speeds, versus the standard handcontroller's four speeds. Meade also announced an electronic focuser for

Figure 1.4. ETX-90EC. Photo courtesy of Meade Instruments Corp.

the ETX that you control from the standard handcontroller or the Autostar handcontroller. I will discuss the Autostar in more detail later in this chapter.

The ETX-90EC retained the suggested $595 retail price, a bargain considering that third-party add-ons to bring the same capabilities to the original ETX were available for several hundreds of dollars. The Autostar was priced at $149. The combination ETX/Autostar made for an amazingly low-priced GOTO telescope system, and the popularity of the ETX increased even more.

ETX-125EC

If a 90 mm (3.5-inch) aperture GOTO telescope is popular, having a 5-inch aperture with its increased "light gathering" power (about twice that of the 90 mm ETX) would be even more popular, or so Meade must have

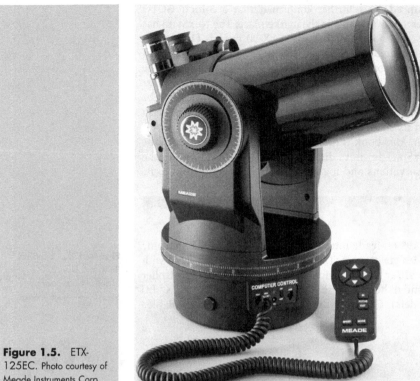

Figure 1.5. ETX-125EC. Photo courtesy of Meade Instruments Corp.

decided (correctly, as it turned out). In May 1999, they announced the "ETX-125EC" telescope that could use the same optional #497 Autostar controller. This 127 mm (5-inch) aperture telescope was a scaled up ETX design but could use most of the same accessories as the "-90" models. The ETX-125EC also has a longer focal length (1900 mm vs. 1250 mm) that, in combination with its larger aperture, allows for the use of higher magnifications than on the ETX-90 models. The ETX-125EC includes an 8 × 25 mm right-angle finderscope. The design is still relatively compact and is almost as portable as the earlier ETX models. Upon its initial release, there were reports of some problems, including large image shifts when focusing and poor optical collimation. Meade identified the cause and fixed the problems.

The ETX-125EC is priced at a suggested retail price of $895 with the Autostar still at $149. Even with the initial glitches that the ETX-125EC suffered, it is still a very popular telescope. Shortly after it was released, a competing product, the Celestron NexStar5, was announced. This further confirmed that a 5-inch GOTO telescope was viable in the marketplace. The NexStar5 has the same focal length as the ETX-90 models (1250 mm) and includes a computerized GOTO handcontroller as standard but at a higher system cost ($1195 versus the ETX-125EC/Autostar combination at $1044). The NexStar computer was not user-upgradeable but had to be returned to the factory for updating. The NexStar5 also had its share of "new product" problems.

As this book is mostly about the ETX-90 models, the NexStar is not discussed here even though the observations and many of the tips are applicable to it.

DS Models

In 1999, Meade introduced a line of "Digital Electronic Series" telescopes. These telescopes come standard with an alt/az mount tripod and an electronic controller (similar to the one that comes with the ETX EC models). Currently the following models are available:

Figure 1.6. DS-60EC (left) and DS-127EC (right). Photos courtesy of Meade Instruments Corp.

DS-60EC (60 mm, 2.4-inch) refractor for $299
DS-70EC (70 mm, 2.8-inch) refractor for $349
DS-80EC (80 mm, 3.1-inch) refractor for $449
DS-90EC (90 mm, 3.5-inch) refractor for $499
DS-114EC (114 mm, 4.5-inch) reflector for $399
DS-127EC (127 mm, 5-inch) reflector for $449

I mention these DS models because you can add an optional #495 Autostar to them for $99. While many of the observations and some of the tips in this book are applicable, the DS models are not discussed further.

ETX-60AT, ETX-70AT

In June 2000 Meade announced a different series of ETX telescopes. These low-priced 350 mm focal length refractors use the similar fork mount and base design as the ETX-90EC and ETX-125EC. Currently the following models are available:

ETX-60AT (60 mm, 2.4-inch) for $299
ETX-70AT (70 mm, 2.8-inch) for $349

Figure 1.7. ETX-60AT (left) and ETX-70AT (right). Photos courtesy of Meade Instruments Corp.

As with the other ETX models, no tripod is included as a standard accessory but a #494 Autostar is included. In some markets, ETX-60EC and ETX-70EC models were sold. They lacked the Autostar but included a basic handcontroller. The various Autostar models are discussed in more detail later in this chapter. As with

the DS models, much of this book is applicable to the ETX-60AT and ETX-70AT.

LX90 8-inch

Apparently in response to the release of the Celestron NexStar8 (8-inch model, $1899), in June 2000 Meade

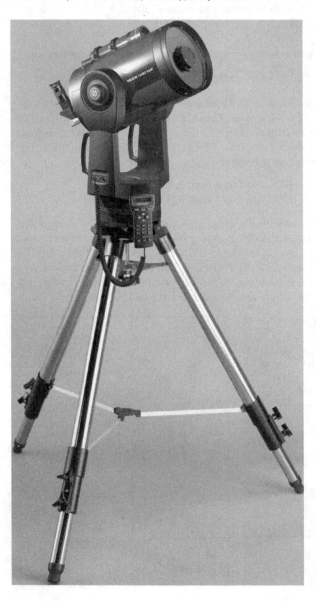

Figure 1.8. LX90 8-inch. Photo courtesy of Meade Instruments Corp.

announced the "8-inch LX90" f/10 Schmidt–Cassegrain telescope for $1695. In this case, the "90" in the name had no relationship to the aperture (203 mm, 8-inch). (Meade already had a very popular line of high-end professional telescopes called the "LX200".) The fork mount and base are similar in design to that of the ETX-125EC. An alt/az mount tripod is included and the Autostar model #497 is standard. Since an 8-inch telescope is not in the realm of "small telescopes", the LX90 is not discussed further, although my discussions about the Autostar will be applicable.

Autostar

Figure 1.9. Autostar Controller. Photo courtesy of Meade Instruments Corp.

The original Autostar was announced with the ETX-90EC model in January 1999. It generated a lot of interest in the amateur astronomy world as a low-cost GOTO computer, allowing the user to select an object from its large database of objects and have the ETX-90EC automatically locate the object in the sky and move the telescope to point at it. Once set up for the user's location, current date and time, and mounting method, the Autostar will track objects whether the ETX is mounted in equatorial mode (ETX base tilted to match the viewer's latitude) or mounted in altitude/azimuth mode (base horizontal). The user can add objects to the Autostar database and even update the Autostar software itself. Since first being introduced, the number of Autostar models has increased and new versions of the software, with up to 30 000 objects, have been released on Meade's Web Site.

The following table compares each of the currently available models

	Autostar #497	Autostar #495	Autostar #494
Compatible telescopes	ETX-90EC ETX-125EC DS models 8-inch LX90	DS Models	ETX-60AT ETX-70AT
Number of objects in database	30 223 (with latest software; originally it was 14 000)	1500	1450
Price (US$)	$149 (Included with LX90)	$99	Included with these models

While using an Autostar is not required to learn and enjoy the night sky, having one is worthwhile for those times when you want the convenience of a GOTO system.

Accessories

When you purchase an ETX (or most any telescope), it usually comes with a standard set of accessories: one or more eyepieces, a finderscope, and perhaps a carrying case. The actual accessories included may be dependent upon where you purchase the telescope and what the dealer provides as part of the price. These accessories are enough to let you begin enjoying your telescope. But, as with many hobbies, you will want to add more over time and you will likely spend more on accessories than you did for the basic telescope. In this section I discuss some useful accessories that will increase your enjoyment of visual astronomy. Accessories for photography are discussed in Chapter 5.

Eyepieces

Typically, one of the first additional accessories most telescope users will add is another eyepiece. Eyepieces come in various focal lengths, fields-of-view, and optical designs. For most ETX users, only the first two are really important; but using high quality eyepieces will allow your ETX to perform at its best. Eyepiece focal length is important since, as mentioned earlier, it is the indicator of the "power" of the eyepiece. But this does not always mean that you will see more with shorter focal length eyepieces. As the power is increased, the amount of light being passed through the eyepiece decreases, dimming the image you see. Also, at higher magnifications the image begins to "break up" or become fuzzy. There is a theoretical maximum power for any telescope that can be calculated by this formula:

Max magnification = 60 × Telescope aperture in inches

For the ETX-90EC, which has a 90 mm or 3.5-inch aperture, this formula yields a maximum magnification of 210×. A more conservative calculation is:

Max magnification = 2 × Telescope aperture in mm

Doubling the 90 mm value yields 180× for the ETX-90. Since most telescope apertures are given in millimeters (mm) today, doubling the aperture is an easy way to estimate the maximum theoretical magnification.

Note that I said "theoretical magnification". The ETX optical quality is very high. If you use expensive, high-quality eyepieces, you can easily exceed the theoretical maximum on bright objects, such as the Moon and some planets. Viewing faint nonstellar objects, such as nebulae and galaxies, will not be helped by using the highest possible magnification, as the image will appear too dim and fuzzy, even to the point of being invisible. Using lower powers will provide the best views of faint extended objects.

Another factor to consider when selecting eyepieces is that with "normal" eyepieces, as the focal length decreases, the field-of-view (FOV) decreases. The FOV is the amount of the sky (or the Moon's surface) that you can see without moving the telescope around. Usually, normal eyepieces have a perfectly acceptable FOV, but there are times when seeing a wider expanse of the sky is nice. For example, when looking at some open star clusters or the nebula-rich central area of the Milky Way, you might enjoy the view more if you could see more sky through the eyepiece. If you want to do "comet hunting" in the hopes of discovering a new comet, using a wide field-of-view eyepiece is almost mandatory. Eyepieces that provide wider fields-of-view typically come in two varieties: wide angle (WA) and ultra-wide angle (UWA). For a given magnification, cost typically increases as the FOV increases, since optical design and quality are factors that determine whether the field is "flat" (i.e., not distorted) and in focus across the whole field-of-view. If having a wide FOV is important to you, definitely look for a high-quality WA or UWA eyepiece.

The ETX-90EC includes a 26 mm eyepiece providing 48× or a magnification of objects of 48 times. To calculate the magnification that an eyepiece will yield with any telescope, use this formula:

Magnification = Telescope focal length ÷ Eyepiece focal length

Inserting the values for the ETX-90EC and its included eyepiece you get:

Magnification = 1250 mm ÷ 26 mm = 48×

With this eyepiece, the Full Moon (approximately 0.5 degrees in apparent diameter) nearly fills the eyepiece field-of-view. All the visible planets will appear as very small disks ("pinhead" or slightly larger or smaller) with this eyepiece. Very quickly you will want more magnification (or power) to see details of the craters on the Moon and to get better views of the rings of Saturn or cloud bands on Jupiter.

A good second eyepiece is the Meade 9.7 mm, or something of a similar focal length, providing 128× on the ETX-90. At this magnification, details on the Moon will be quite visible and Saturn and Jupiter will be breathtaking. However, there are times when you may want less power, as when viewing star clusters. I acquired a Scopetronix 40 mm (31×), which is an inexpensive low-power eyepiece yielding a wide field-of-view. Rather than having several eyepieces, each with a different focal length, some users prefer a "zoom" eyepiece. Zoom eyepieces provide a variable range of focal lengths, such as from 8 mm to 26 mm. Alternatively, if you have a small set of eyepieces, a very useful accessory is a "Barlow lens". These typically double (known as "2×") or triple ("3×") the magnification of the eyepiece that they are used with. There is even a variable magnification Barlow lens available going up to 5×. Using a 2× Barlow lens with the 26 mm eyepiece on the ETX-90EC provides 96× (or twice the normal 48×). I like the 2× Barlow lens for use with my set of eyepieces, as it essentially doubles my effective number of eyepieces.

When considering eyepieces it may be important to remember that usually the "eye relief" decreases with shorter focal length eyepieces. Eye relief is how far behind the eyepiece the image focuses and therefore the distance at which your eye is placed. Short eye relief eyepieces can be a problem for those users who wear eyeglasses to see clearly or can even collect debris-attracting oil from eyelashes.

Just as the Barlow lens increases the magnification of any eyepiece, a "focal reducer" or "wide field adapter" will decrease the magnification and increase the FOV. This is a relatively inexpensive way to further expand the range of eyepieces that you have but at the expense of image quality, since some distortion and focus problems are inherent in these.

Optical accessory prices vary widely, depending upon the design, focal length, FOV, and manufacturer. Expect to spend from $50 to several hundreds of dollars per item.

Mounts

For many users, setting the ETX on its tabletop tripod legs or even on any flat stable surface works fine. But for others or for those times when there is no convenient flat surface at a useable height, having a tripod really helps. There are many to choose from, such as a Bogen (or Manfrotto) sturdy photographic tripod, a Jim's Mobile Inc. (JMI) "Wedgepod" or "Megapod", or even Meade's tripods. A key factor when considering a tripod for your ETX is how sturdy and stable a platform it is. While the ETX-90 is not heavy, it does cost several hundred dollars and you do not want the tripod to collapse or tip over. In addition, especially when using higher magnifications, any vibration in the tripod will cause the image you are viewing or photographing to bounce around. Even slight breezes or just trying to focus the image can cause image motion, so the sturdier the mount the better. Even the surface on which the tripod is sitting can affect the degree of vibrations; grass or a soft surface will dampen out vibrations faster than a hard surface like concrete. Some users place the tripod leg tips on foam padding or other soft material to reduce vibrations.

When using the ETX in alt/az mode, almost any sturdy tripod will work (if you have the proper adapter). It is when the ETX is mounted in polar mode (equatorial mounting) that stability problems can become evident with some photographic tripods. At low latitudes, the ETX has to be tilted a lot for its vertical rotation axis to be parallel to the Earth's rotational axis, placing the ETX center of gravity well to the side of the center of balance of the tripod. This creates an unstable condition, putting your ETX and wallet in jeopardy. A better choice is to use an equatorial mount tripod or a "wedge". These allow the ETX to be tilted to match your latitude while keeping it centered over the tripod. Some wedges even have holes to hold eyepieces or other accessories, making them quickly and easily accessible.

Tripods and wedges will cost from $100 and up, with the better ones costing upwards of $300.

Finderscopes

Many users find the standard ETX-90 finderscope limiting. It has a small field-of-view, a thick cross-hair, and is straight through as opposed to a right-angle finder (which is standard with the ETX-125EC model), which some users prefer. There are times when having a different type of finderscope is useful. For example, when the ETX is pointed high in the sky, you have to bend way over to look through the straight-through finderscope. This can be awkward, so having a right-angle finderscope helps. The problem with right-angle finders is that your eye is not actually "looking" in the direction of the sky that the telescope is pointed at but perpendicular to this direction. Some users find this disorienting and so prefer to look through a finderscope in the direction that is parallel to the line-of-sight of the telescope. For most users, a "red-dot" or "1×" (no magnification) style finderscope makes a great addition to the ETX. Many are small, such as the Scopetronix LightSight, which projects a red-dot against the sky background. Some finderscopes are larger but have other features. The Rigel Systems QuikFinder, which is a 1× finderscope, has an illuminated reticle to assist in locating objects. Both the red-dot and 1× finderscopes have wide fields-of-view, and you look at the sky in the direction in which the telescope is pointed. These nonmagnifying finderscopes work best with bright objects and will get the telescope pointed close enough to the object to use a finderscope with some magnification. For locating fainter objects or for when you need more precise pointing, you need a finderscope with some magnification, such as the 8 × 21 mm finderscope that is included with the ETX-90EC. If you prefer a right-angle finderscope, there are modification kits for the 8 × 21 finder or you can replace it with a complete right-angle finderscope. The decision to add another finderscope or replace the existing finderscope is yours, depending upon how you use the ETX. My ETX-90RA has the standard 8 × 21 mm finderscope converted to a right-angle with the Shutan Camera and Video (Apogee) Right-Angle Adapter, as well as the Scopetronix LightSight. I also use the Rigel Systems QuikFinder when doing piggyback astrophotography (described in Chapter 5). Prices range from $25 to $50 and up, depending upon the design.

Filters

Colored glass filters enhance the image you see by allowing only specific light frequencies to pass through. They increase the contrast of some portions of the image. Selecting the right filter can dramatically improve the view of the object. Good-quality filters, meaning they are optically "flat" (i.e., do not have distortions) and have uniform color, typically cost $10–20. Filters come in various densities, meaning that some pass more or less light. Dark filters cannot be used with the ETX-90 models (or other small telescopes), as there is insufficient light reaching the filter due to the small aperture of the telescope. Filters are identified by a "Wratten" number. Some filters I recommend for the small telescope user are:

> #8 Light yellow: improves views of Martian maria, some Jupiter/Saturn cloud bands
>
> #11 Yellow-green: improves views of Martian maria and polar ice caps, some Jupiter/Saturn cloud bands
>
> #12 Yellow: improves views of Martian maria, some Jupiter/Saturn cloud bands
>
> #15 Deep yellow: improves views of Martian maria and polar ice caps, some Jupiter/Saturn cloud bands
>
> #21 Orange: improves views of the Martian surface, some Jupiter/Saturn cloud bands
>
> #56 Light green: improves views of Martian polar caps and some cloud formations, some Jupiter/Saturn cloud bands
>
> #82 Light blue: improves some Jupiter/Saturn cloud bands

If you observe from an area with considerable "light pollution", one of the broadband or narrowband nebula or "light pollution" filters may be helpful. These either block certain wavelengths (or "colors", if you prefer a nontechnical term), typically emitted by street lighting, or pass certain wavelengths, typically colors emitted by emission nebulae. I have used the Celestron Light Pollution Reduction (LPR) and was pleased with the results. It blocks mercury and high and low pressure sodium vapor emissions and the natural sky glow due to neutral oxygen emission in the atmosphere. Using this filter results in improved views by increasing the contrast of the object against the sky background. According to the documentation supplied with the filter, the LPR passes over 90% of the desirable wavelengths, which improves views of emission neb-ulae. With a small telescope, you are limited to lower magnifications, as the filter is rather dark. On brighter objects, it does effectively increase the contrast of the object over the surrounding sky, which can make some details stand out. These types of filters are more expensive, typically $70–100.

A Moon filter can be compared to wearing sunglasses during the day to reduce glare and make seeing easier on the eyes. A Moon filter can definitely increase your enjoyment of studying our closest neighbor even with small telescopes. A Moon filter costs about $15.

A solar filter makes a nice accessory when used with caution. Viewing the Sun with the ETX requires a solar filter. Whereas some telescopes can be used to project the Sun's image onto a white surface for viewing, the plastics used in the ETX will melt if unfiltered sunlight enters the tube! Obviously, you want a high-quality solar filter from a reputable manufacturer to avoid putting your eyesight and equipment at risk. The filters discussed previously screw onto an eyepiece. While there are solar filters made for use on an eyepiece, I strongly advise against using them as they are prone to failure from heat buildup. If the filter cracks while you are looking at the Sun, say goodbye to sight in that eye! The best solar filter is a "full aperture" one that covers the aperture end of the telescope to prevent damaging heat buildup inside the telescope tube. Although you can make your own solar filter using appropriate material (e.g., solar filter quality Mylar or the Baader Planetarium AstroSolar Safety Film), many users prefer to buy a ready-made one, costing about $60–100.

Antidew Devices

If you have ever gone outside on a clear night and noticed that the grass or your lawn chairs are wet, you may have a dew problem. Obviously, you do not want dew collecting on your telescope, especially on the optics. A simple device that can help avoid this is a "dew shield" or "dew cap". Several companies make these for the ETX (and other telescopes), or you can easily make one yourself from thin cardboard rolled into a tube or a better one from large-diameter lightweight PVC pipe tubing. Paint the tube flat (not glossy) black and slide it onto the sky end of the telescope with at least enough of the tube extending beyond the end of the telescope tube to equal the aperture diameter (90 mm on the ETX-90). If you experience heavy dewing, you may need more extension. Alternatively, you can purchase (or make, see my ETX Site for details) a heater for your telescope that keeps dew from forming on the telescope optics. Costs range from about $25 to around $100, depending upon the size and design. I do not recommend using a hair dryer to remove dew, as the heat may damage ETX components.

Carrying Cases

An accessory that some users will need to buy (or make) is a telescope carrying case. Users who only take the telescope outside on their balcony or backyard and like to display the telescope indoors when not in use, which is typical with the ETX given its beauty, will likely never need a case for the telescope. However, users who travel to locations with dark skies or take their telescope with them on vacation trips will need some type of carrying case to protect the ETX and hold some accessories. There are two kinds: hard-sided and soft-sided. Hard-sided cases offer more protection and are usually more expensive than soft-sided cases. If you travel by commercial airline, be aware that some hard-sided cases will not qualify as carry-on luggage. Checking your telescope, even in the best hard-sided case, is not a good idea. Sometimes the rough treatment the case will receive can knock the telescope out of collimation, or worse. I purchased the Shutan Camera and Video "Deluxe Softsided Case" with the Backpack Option for my ETX-90RA and used it when I traveled to Australia in 1999. This case, packed with the ETX, eyepieces, other accessories, and a 35 mm camera, easily fits into the overhead bins or under the seat of commercial airliners.

Cases for eyepieces and accessories can also be handy. I found that zippered cosmetic bags make excellent holders for these. Others like to use sealable plastic food containers. Protecting your telescope and its accessories from dust and damage is a good habit to get into. Good cases will cost from $50 to $200, depending upon the size of the case and whether there are foam-packing materials included.

Books and Charts

Learning your way around the night sky will increase

your enjoyment of using the ETX (or any telescope). If you have the Autostar, you will not have to spend time looking up locations of objects to view. However, you will still need to know some basics of what is where in the sky. This is when even the most basic of sky charts comes in handy. Whether you have a GOTO telescope or not, there will be times when you need more detailed charts that show fainter stars and objects. The monthly astronomy magazines such as *Sky & Telescope* and *Astronomy* are excellent sources of charts depicting what is visible for each month. Also, many astronomy-related Web Sites and books have this information. See the Appendix for more on these.

Software

Astronomy software is an excellent source of nightly sky charts. There are many programs available for Windows, Mac OS, Palm OS, and WindowsCE, some of which can even be used with the Autostar to select and view objects through the ETX. Most programs can be configured for your location, date, and time, thereby giving you an accurate presentation of the sky. Some even have the ability to make up and print observing lists with lots of details on your selected objects, including "finder charts" to help you locate the objects. See the Appendix for more on software.

If you have an Autostar and a desktop computer, you will likely want to update its software as Meade releases new versions. To do that, you will need Web access to download the latest Autostar Updater and "ROM" files for your Autostar, and a serial cable to connect your computer to the Autostar. There are several kits available; be certain to purchase the one for your model Autostar. They cost $20 and up, depending upon the source and model. If you have a Macintosh or USB-only computer, you will need to purchase the appropriate

Other Add-ons

serial adapter for your computer.
The original model ETX only had a motorized drive for the right ascension axis. To engage the drive, you manually locked the RA axis. To move the telescope to locate objects, you unlocked both the RA and

declination axes and manually slewed the ETX. This worked well for many users but was inconvenient when slewing the scope around small portions of the sky, such as when "star hopping" (see Chapter 3). JMI was the first to have an ETX add-on that provided electronic control over the DEC axis. The JMI "MotoDec" added a small motor to the outside of the fork mount that could rotate the manual DEC control knob. Its handcontroller contained the battery and buttons for slewing. This solution worked well for movements in declination, but some users wanted more. Along came Scopetronix and its "Microstar Dual Axis Drive Corrector" and its successor the "Microstar II+". The Microstar used a replacement ETX electronic circuit board to accomplish dual axis control. Installation was more involved than with the MotoDec since you had to open the ETX base and remove the original circuit board and insert the Microstar board. While this did void any warranty on the ETX, it was actually an easy installation. The electronics controlled the existing RA drive motor while an external motor (similar to the MotoDec motor) was attached on the outside of the fork mount. The handcontroller provided electronic control over both axes. Using the Microstar made the ETX feel like a bigger scope with full electronic control over slewing and tracking. Such add-ons will cost you $100–200 and up.

An inconvenience (some would say "annoyance") is the small focus knob at the rear of all ETX models. It is difficult to focus the image without inducing a lot of vibration, especially at high magnifications. One of the first ETX tips to gain widespread acceptance was the famous free focusing aid: clip a clothespin to the focus knob. JMI had the first electronic focus modification, called the "MotoFocus", that added a small motor and handcontroller (or it could use the same controller as the MotoDec) to control focusing. This worked well and almost completely eliminated vibrations, as your hand was not actually touching the ETX when focusing. Scopetronix came out with a similar modification that worked in conjunction with their Microstar II+. The Microstar II+ also supported the JMI MotoFocus. When the ETX-90EC was released, Meade also offered an optional electronic focuser that would work with or without an Autostar. Nonmotorized focusing aids are also available, including larger replacement knobs and the Scopetronix "FlexiFocus", which added a flexible cable between the ETX and your hand, minimizing hand-induced vibrations. These items cost from $25 or less (for manual focusing) to about $100 (electronic focusing).

As described at the beginning of this chapter, the ETX design makes objects appear reversed left-to-right. This effect is acceptable for astronomical use but can be unacceptable for terrestrial use. Adding a "right-angle erecting prism" to the rear port of the ETX will correct the orientation of the image. Unlike an eyepiece in the normal position on the top of the ETX, this attachment can be rotated, making the eyepiece position more convenient for some users. Adding a "star diagonal", which also attaches to the rear port, will put the eyepiece at a more convenient location for some users for astronomical use. I like to have an eyepiece in both locations. I then switch between eyepieces using the flip-mirror on the ETX-90RA. You will spend $50–100 for these accessories.

This chapter has described some basics of the ETX (and some other telescopes) and a few of the accessories and add-ons that can enhance your use and enjoyment of the ETX. The world of the ETX is a wide and still expanding one. But it is not a perfect world. Sometimes things can go wrong or the documentation supplied may not be as clear to you as it should be. The next chapter will discuss some hints and tips that may help.

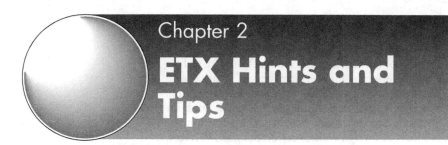

Chapter 2
ETX Hints and Tips

Important note: Some topics discussed in this chapter may invalidate your telescope's warranty or not be applicable to your telescope model. Neither the publisher nor the author is responsible for any damage or problems that may result from using the information presented. The information presented in this chapter is from my personal experience or from the personal experiences of other ETX users as posted on my ETX Web Site (http://www.weasner.com/etx).

Whether you have purchased a new or used telescope or have just used your telescope for several years, there may come a time when you feel that it is just not performing the way it should. Or you would like to get more use of out it. While most of this chapter is for the ETX models, some information may be applicable to other telescopes or will at least start you thinking about what you might be able to do with another model.

There are three areas where you may be able to improve your telescope's performance: optical, mechanical, and software. I will discuss these followed by some other tips. See my Site for even more hints and tips.

Optical

Cleaning Optics

Most users think that perfectly clean optical surfaces are required for the best views. While that is true, there is no need to go overboard and clean the optics before every use. In fact, if you have provided your telescope with some protections during storage and usage, there is usually no need to clean the optics unless they get really dirty or smudged. Small amounts of dust or hair will not deteriorate the performance to a noticeable extent. To determine whether the optics need cleaning, shine a strong light at, not into, the telescope. Look for fingerprints, smears, areas of dried condensation, or large pieces of foreign substances on the glass surfaces. If you see dust or hair, do not try to blow it off with your mouth; you will likely end up having to clean the spit off! Use a can of compressed air (follow the instructions to avoid getting moisture on the optical surfaces) or a squeeze-type blower to blow any debris off. This should always be your first method for dealing with dirt, as it avoids touching the optical surfaces. These are available from electronics and camera stores. If the foreign objects cannot be blown off, obtain a good-quality soft camelhair type brush (available at camera stores). With minimum pressure brush the dirt away in strokes towards the edge of the optical surface.

To remove smears or dried condensation, it will be necessary to contact the glass or mirrored surface. This requires extra care, especially with a mirror. Cleaning materials available at your local camera store can help. DO NOT use household window cleaning solutions on your telescope optics, as you might inadvertently remove the special coatings on the optics and deteriorate their performance. I think the best solution for dealing with small areas, or small surfaces such as an eyepiece, is to use one of the LensPen products available at camera and telescope dealers. A LensPen ($10) can effectively and easily clean even the most stubborn of substances off your optics. They come in two sizes: one suitable for large surfaces such as an objective or corrector lens, and a smaller one suitable for cleaning an eyepiece.

If more serious cleaning is required, for example to clean that telescope that was stored uncovered in a shed for several years, you can use a highly diluted soap solution (liquid, not solid) and a soft lint-free cloth. Again, use as little pressure as necessary to soak and remove any accumulated grime. Dry using another lint-free cloth by touching the cloth to the surface but without making any swirling or wiping movements. Finally let the optics air dry in a protected or covered area. Remove any remaining water spots with a LensPen.

Collimation Tests

With some telescopes, the optics can become misaligned, either due to rough handling or just from normal flexing during use over long periods of time. Optics that are out of alignment will not perform as they should, resulting in distorted or blurry images. Some telescopes can be easily collimated by hand and some cannot. Some telescopes, typically refractors, should never need realigning. For those that do and are meant to be user-collimated, follow the instructions in the telescope manual. The ETX-90RA, ETX-90EC, and ETX-125EC are not supposed to get out of collimation in normal usage and are not intended to be collimated except by the factory. However, the reality is that the ETX Maksutov–Cassegrain models can get out of collimation and can be user-collimated. There is one situation that users should not attempt to correct: if you notice the secondary mirror, which is mounted on the inside of the corrector lens, seems to have slipped off-center, contact Meade to have it repaired.

So, how do you know your telescope needs to be recollimated? Usually, the first clue is that views are not as sharp as they used to be (but you have ruled out dirty optics as the culprit) or you see a very large shift in the image position across the eyepiece FOV as you turn the focus knob. To see if your telescope is out of collimation, do a simple "star test". Center and focus on a bright star using a moderate power eyepiece (the supplied 26 mm eyepiece works fine on the ETX models). Then move the focus to either side of the infocus position. As you move further out of focus on either side of the infocus position, you should see concentric rings of starlight appear (as shown in Figure 2.1; the small circles are from dust on the optics).

If the rings are circular or nearly so, there is no need to adjust the collimation. However, if the rings are oval, you may wish to try collimating your telescope. Collimating an ETX is not something that everyone should attempt. It requires a great deal of patience and some practice to get it right and not make the telescope perform worse than it did before trying to collimate it. You will likely need to disassemble and reassemble your ETX multiple times over several hours before you either get it right or you give up in frustration and send it back to Meade for their expert

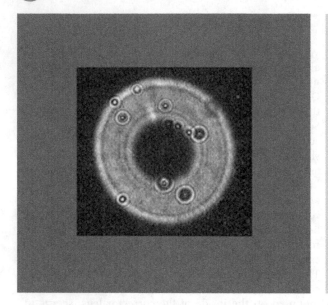

Figure 2.1.
Concentric rings.

collimation. While the actual process is beyond the scope of this book, if you want to give it a try, the ETX Maksutov–Cassegrain models can be collimated by following the steps documented on my ETX Web Site. Please note that Meade may change their designs and invalidate those instructions. If in doubt about anything in the process, do not try it yourself; send the telescope back to Meade.

Blurry Images

If the image you see looks out of focus or blurry, there may be one of two possible reasons. One is addressed in the "Tune-ups" section later in this chapter. The other is that sometimes one or more of the pieces of glass in an eyepiece can shift. This can occur if a filter has become stuck and while attempting to unscrew it from the eyepiece tube, you actually loosened the tube instead. You may or may not hear a "rattle" when you lightly shake the eyepiece. Many times all that is required is to hold the eyepiece upside down, gently tap on the side of the eyepiece tube, and retighten the tube that was loosened. In some cases, the eyepiece may be so badly out of adjustment or assembled incorrectly that it needs to be replaced.

Mechanical

Tune-ups

There are three simple tune-ups that any user can and may need to perform. These require no disassembly of the ETX, but you should still exercise some caution when doing them.

The first is adjusting the declination setting circle to read correctly. When the ETX is set up in altitude/ azimuth (alt/az) mode with the base horizontal, the DEC circle should read 0 degrees when the OTA is horizontal. If it does not, you can easily fix this. Using a spirit level, first level the base and then the telescope tube. Next, slightly loosen the DEC scale knob. When the knob is loose enough (but before it actually comes off), you should be able to rotate the setting circle by moving it with a finger on its surface. Rotate the circle until it shows 0 degrees beneath the triangle pointer at the top of the fork arm. While holding the setting circle in this position with a finger, retighten the DEC scale knob. Verify that the OTA and base are still horizontal and that the declination setting circle shows 0 degrees. Repeat if necessary.

By the way, if you use the setting circles (and many users do not, especially those with an Autostar), you may find it hard to see the triangles in the dark, even with your red-covered flashlight. You can make it easier to see the small triangle pointer on both axes by adding a small drop of luminescent paint (available at some photography or art supply dealers) to the triangle. Your flashlight will charge the pointer while you are looking at the circles.

The next adjustment requires a tool (a small Allen wrench or hex key) but is rarely required. Meade properly warn against overtightening the axis locks, but sometimes you may feel that you cannot get a good strong lock on one or both of the axes. If the right ascension axis lock needs more tension, all that may be required is to reposition the RA lock knob (original ETX) or lever (ETX-90EC, ETX-125EC). To do this:

1. Move the RA lock into its fully locked position.
2. Loosen the setscrew on the side of the knob or lever.
3. For the knob, lift it up a little bit on the shaft and retighten the setscrew. For the lever, lift it off the shaft, reposition it more to the left, and retighten the setscrew.

In making this adjustment, you allow more movement of the shaft, which can result in a tighter axis lock. Again, use caution and do not overtighten the lock or you can damage it. If the DEC axis is not adequately locking, more serious surgery is required (beyond the scope of this book but documented on my ETX Web Site).

This next tune-up is simple, and the need for it is usually noticed (if at all) when you get more eyepieces. You may notice that with one or more of those new eyepieces, you just cannot quite get a good infocus image. If the focus knob is reaching the ETX backplate before focusing the eyepiece, it is necessary to slightly adjust the knob position.

1. Point the ETX slightly upward to prevent the focus shaft from slipping inside the OTA (which can be a real pain if it occurs).
2. Using an appropriate size Allen wrench (or hex key), loosen the setscrew on the side of the focus knob.
3. Slide the knob slightly outward on the shaft (away from the backplate).
4. Retighten the setscrew.

Verify that the eyepiece now focuses. If not, repeat the process. If you do get it to focus, you should be OK with all your eyepieces. If you still cannot reach a focus, it may be that the eyepiece is assembled incorrectly (see the "Blurry images" tip earlier in this chapter).

Speaking of focusing eyepieces, as users add more eyepieces to their collection, they quickly notice that each eyepiece has a different focus setting. That is because most eyepieces purchased are not "parfocal" (having the same focus position) with other eyepieces. For some eyepieces, it is possible to make them parfocal with another eyepiece by doing the following.

1. Focus with your most used eyepiece and then remove it.
2. Slide your next most used eyepiece into the eyepiece holder, but do not adjust the focus knob.
3. Slide the eyepiece in and out to see if you can find a position where the object is in focus.
4. If you can reach a focus by moving the eyepiece, carefully make a mark on the eyepiece tube at the edge of the eyepiece holder.
5. Remove the eyepiece and apply some thick or plastic tape around the tube above the mark you made. This

tape will keep the eyepiece from sliding all the way into the eyepiece holder and will make the eyepiece parfocal with the first eyepiece.

6. Repeat with your other eyepieces to see if you can make more parfocal.

You may want to start with a short focal length eyepiece and proceed to the longer focal length eyepieces to achieve the goal of making them parfocal. Note that this will not work with all eyepieces, and so it may not work with any that you have in your collection.

Some owners of the original ETX noticed that their telescope became somewhat wobbly on the base with use. The problem is that some teflon pads on the azimuth (right ascension) bearings inside the base had come off or moved. While this problem has not affected the later ETX models, if you have one of the original ETX telescopes, either new or used, you may ultimately experience this problem. The best solution is to return it to Meade for repair. But if you are adventurous and somewhat mechanically inclined, you can repair this yourself by following these steps.

1. Remove the base cover plate (as you do when changing the batteries).
2. Remove the screw in the center of the base and lift off the fork mount assembly. Use caution not to damage the RA setting circle paper tape.
3. You should see three small teflon pads that the fork mount assembly was riding on. One or more may be out of place (if they have slipped), worn, or missing. Reposition any that are of out of place and replace any worn or missing ones using strong glue. Exact replacements can be ordered from Meade, or you can use substitute materials as long as they can be made the right thickness and are sufficiently slippery.
4. Reassemble the ETX base.

Occasionally, owners of new ETX models have noted that the motorized drives will not slew the telescope in one or both axes when they first try to use their new scope. Sometimes the cause is simple. The grease on the gears may be stiff or unevenly distributed. Unlock the affected axis and manually move the OTA through several rotations in both directions (if the problem is with the azimuth axis) or up and down (if the problem is with the altitude axis) through its full range. If this does not clear the problem, it may be more serious, requiring a return to the dealer or Meade for an exchange or repair. But before you do that, if you can, try a different handcontroller or Autostar to see if that could be the source of the problem.

Other tune-ups require even more surgery by going into the guts of your telescope. Some are for improving the DEC lock mechanism or improving the tracking. These are beyond the scope of this book but all are documented on my ETX Web Site (http://www.weasner.com/etx) in the Telescope Tech Tips section.

Polar Aligning without Autostar

If you have a telescope without alt/az tracking capabilities such as provided by the Autostar, then you must set up the telescope in an equatorial or polar mounting mode if you want it to track celestial objects as they move westward across the sky due to the Earth's rotation. This allows only one axis and drive motor, right ascension, to be used in the tracking.

There are times when you just want a quick alignment since accuracy may not be important for that observing session. Many times all you need to do is point the telescope polar axis towards the north (or south), not worrying about things such as putting Polaris in an eyepiece, leveling the tripod base, or properly setting the latitude. Tracking errors will be worse for really sloppy set-ups, better for more careful (or lucky) set-ups. But if you just want to view one or two objects and only for a few minutes, this technique will likely be good enough. In fact, there are times when just setting up in alt/az mode works well. Say you just want to take a quick look at the Moon at a low power. Why take longer to set up than you will actually observe? If the telescope is already in alt/az mount mode, just take the telescope outside, point the telescope at the Moon, and look. Enjoy the experience, not what it takes to get there.

For those times when you want more accuracy in the tracking, such as to follow a planet for several minutes while you make a drawing at the eyepiece, you have to do a better polar alignment. You do this by tilting the telescope vertical rotational axis (the axis perpendicular to the horizontal base) to compensate for your observing location's latitude. Your telescope manual probably has a good description of this. If you are at the equator, the axis will be parallel to the ground (a really difficult to mount situation); if you are at one of the Earth's poles, the axis will be perpendicular to the ground (the same as an altitude/azimuth mounting mode). If you are someplace between these two extremes, the tilt is 90 degrees minus your latitude. If you are at 30 degrees latitude, the telescope is tilted over 60 degrees from the vertical. Or you can think of it as being tilted up from the horizontal by the same amount as your latitude. At 30 degrees latitude, you would raise the vertical axis from horizontal by 30 degrees. Be aware that most consumer grade tripods or wedges have very imprecise latitude markings and you need to ensure that the tripod base is perfectly level for the tilt to be accurate. Once the telescope is properly tilted, you must point the polar axis at the north (in the northern hemisphere) or south (southern hemisphere) Pole. In the northern hemisphere, the bright star Polaris is close to the proper position and most small telescope users use it when it is visible; just center Polaris in a low-power eyepiece. There is a faint star near the south Celestial Pole. It is Sigma Octans (also known as Polaris Australis). It is a Magnitude 5.5 star located at 21h 48m 46.7s and 88d 57m 23s. When these stars are not visible or if you are in the southern hemisphere, it is acceptable to use a magnetic compass to get the proper direction. However, be aware that many locations on the Earth have a "magnetic variation" (the difference between True North and Magnetic North) of several degrees, which when large can cause tracking errors in a very short period of time. If you are using a compass, determine your local magnetic variation first. A nearby airport is the most accurate source of this information. Also, keep in mind that there can be many other influences on the accuracy of a magnetic compass. It may not be the best method for your location. The result of the tilt and pointing is to get the telescope vertical rotational axis to be parallel to the Earth's rotational axis. At this point, the alignment is imprecise but probably good enough for most visual purposes. You can now find objects in the sky based upon their right ascension and declination values, and the right ascension drive (electric or manual) will track the objects reasonably well over short periods of time as they move in the sky. If you always observe from the same location, you may want to mark the positions of your tripod legs on the ground to facilitate future set-ups.

There may be times when even the above alignment is not good enough for what you plan to do with the telescope. If you want to attempt long-duration astrophotography or just want the object to stay put in that very high-power eyepiece; you will need to refine the accuracy of the alignment. A common technique for this refinement is called the "declination drift method". This can be very time-consuming since you have to allow a star to track across the eyepiece FOV, adjust an axis by slightly raising or lowering one or more tripod legs or otherwise adjusting the polar axis tilt, reorienting the telescope to improve the pointing towards the pole, or usually some combination of these. Once you complete the steps described below, your polar alignment will be very precise. If you are in the southern hemisphere, exchange "north" and "south".

Declination Drift Method

1. Point the ETX fork arms at Polaris.
2. Locate a star approximately due south of you and near the celestial equator. Place this star at the north or south edge of a high-power eyepiece. Make the star's image slightly out-of-focus to make it easier to detect drift. Engage the right ascension tracking motor.
3. Watch the star's movement in declination (north and south). If the star moves south, the fork arms (polar axis) are pointing too far to the east. If it moves north, the forks are pointed too far to the west. Physically pick up the telescope and its mount and reposition the fork arms accordingly. Repeat this step until there is no more drift north or south.
4. Locate a star approximately due east of you and near the celestial equator. Place this out-of-focus star at the north or south edge of the eyepiece. If your eastern sky is obscured, use a star in the western sky.
5. Watch the star's movement in declination (north and south). If the star moves south, the fork arms (polar axis) are pointing too low (i.e., pointing below the pole). If it moves north, the forks are pointed too high. Adjust the tilt of the telescope on its mount to reposition the fork arms accordingly. Repeat this step until there is no more drift north or south. If using a star in the western sky, exchange "too low" and "too high" in the above corrections.
6. Start over with step 2. This allows corrections of errors made when repositioning the telescope.

Polar Aligning with Autostar

There are several modes of Autostar operation and each user will pick a preferred method. Some users like to use polar instead of alt/az mounting. Some users prefer "star hopping" (see Chapter 3) instead of having the Autostar locate objects to be viewed. In all cases you still have to align the Autostar at the start. This section provides some tips on doing that.

While the method of selecting menu items varies in the low-end #494 Autostar from the high-end #495 and #497 models, the usage is essentially the same. When powered on for the first time,

1. Select your location from a large list of cities worldwide. (If the Autostar was configured in a store, the language and location selections may be incorrect. If you cannot read the text because the language is incorrect or if the location appears to be incorrect, follow the steps in the "Resetting the Autostar" in the Software section of this chapter.)

2. Select your telescope model and whether the telescope is mounted in polar (equatorial) mode or in alt/az mode.

3. Select "train the drives" mode, which is the operation that the Autostar uses to learn how the drives and axis encoders respond to movements.

Once your Autostar knows about your location and telescope, you do not have to enter these again, as they are stored in the Autostar's memory. You can change these settings at any time or store your precise location in longitude and latitude. Note: in early versions of the Autostar software, west longitude was positive but was changed in version 1.1m to west being negative.

Retraining the drives is a good thing to do if the Autostar ever starts having any GOTO problems or experiences any other GOTO oddities. Once you have done these initial steps, you only have to enter the date and time and select your star alignment method when you power on the Autostar system.

Using your longitude/latitude, the current date/time, and proper HOME position (described shortly) as known starting points, the Autostar calculates what alignment stars are currently visible in your night sky and selects one. It then slews the telescope to point where it believes that star is located. Usually this is within the field-of-view of the finderscope; but sometimes it may be just outside, depending upon the preciseness of your initial set-up. The brightest star in the finderscope or near where the telescope is pointed is usually the correct star. You then manually slew the telescope until this star is centered and confirm to the Autostar that it is centered by pressing the "ENTER" button. The Autostar selects a second star (when using the more precise Two Star Align method) and you then repeat the process. The Autostar does a calculation on the positions of these two "centered" stars. If they match the actual stars, the Autostar determines that the alignment is successful and displays an "alignment successful" message. If there has been some error and the calculations do not agree with your "centered" stars, the alignment has failed and you start over. In many months of using the Autostar, I have seen this occur only a few times. There can be many reasons why the alignment will fail or, even if successful, why attempting to GOTO objects fails to center the object. See in the "Autostar Usage" section later in this chapter.

For the Autostar to perform at its best, it needs to start from a standard orientation known as the HOME position. There are actually two HOME positions; one for altitude/azimuth mounting and a slightly different one for polar (or equatorial) mounting. Follow the appropriate steps for the way you have your telescope mounted. The more precise you make each step the better the initial results will be in putting the alignment stars at or near the center of the FOV of a properly aligned finderscope, and the sooner you will be observing.

Alt/Az HOME Position

1. Position the ETX and tripod system with the ETX control panel on the west side. This does not have to be exactly due west but should be as close as possible. Note: the control panel on the ETX-60AT and ETX-70AT should be on the east side.

2. Level the ETX base horizontally. It does not have to be exactly level; I usually just eyeball it.

3. Unlock the altitude axis and position the ETX Optical Tube Assembly (OTA) to 0 degrees on the altitude setting circle. The OTA should be horizontal. Relock the altitude axis. If the scale is off, it can be adjusted by loosening the knob, rotating the scale, and retightening the knob (see the tune-up tip on this earlier in this chapter). Since you are not likely to be

using the altitude setting circle when you are using the Autostar, you can defer that adjustment until a more convenient time.

4. Unlock the azimuth lever and slowly rotate the OTA counterclockwise by hand until you reach the stop. Counterclockwise means moving it in a direction going from pointing south to east to north to west. Note: not all models of the ETX line have hard stops; if yours does not, skip this step.

5. Rotate the OTA to point to north. This is a clockwise rotation of approximately 120 degrees or from pointing towards the southwest moving through west to north (hard stop models).

6. Continue with the Autostar alt/az initialization and alignment as described earlier.

Polar HOME Position

1. Position the ETX and tripod system with the ETX control panel on the west side. This does not have to be exactly due west but should be as close as possible.

2. Check that you have your telescope base tilted to correspond to your latitude (see the discussion on polar aligning earlier in this chapter).

3. Unlock the declination (same as altitude) axis and position the ETX Optical Tube Assembly (OTA) to 90 degrees on the DEC setting circle. The OTA should be perpendicular (that is, at a right angle) to the ETX base and the OTA pointed towards the pole position in the sky (near the star Polaris in the northern hemisphere). Relock the axis. If the DEC scale is off, it can be adjusted by loosening the knob, rotating the scale, and retightening the knob. Since you are not likely to be using the DEC setting circle when you are using the Autostar, you can defer this adjustment until a more convenient time.

4. Unlock the right ascension (same as azimuth) lever and slowly rotate the OTA counterclockwise by hand until you reach the stop. Note: not all models of the ETX line have hard stops; if yours does not, skip this step.

5. Rotate the OTA to put the eyepiece on top. This is a clockwise rotation of approximately 120 degrees (hard stop models).

6. Continue with the Autostar polar initialization and alignment as described earlier.

Autostar "Quick Align"

There are times when having the Autostar gets in the way of a quick set-up even though it normally takes only a couple of minutes to align the Autostar. You can perform a "quick" alignment that, while not perfect, may be good enough for your observing session. There are different ways of getting a quick alignment.

In one method, you follow the normal Autostar alignment steps but when asked to manually center the selected alignment star, just assume the proper star is centered and press the ENTER key on the Autostar handcontroller. This allows the Autostar to make its calculations and engage tracking. The better your HOME positioning the more accurate this quick alignment will be. This method also works when you cannot see the selected alignment star due to some obstruction or you want to view Venus during the daytime.

Another method is even faster since it skips the time for the telescope to slew to the alignment stars. However, this method will not work if you need GOTO capabilities since the Autostar does not have any alignment to the sky. In this method you:

1. Press ENTER three times to get past the date, time, daylight savings entries
2. Press MODE once (the display changes to "Setup: Align")
3. Scroll down five times to "Setup: Targets"
4. Press ENTER (display now shows "Current Targets:>Terrestrial")
5. Scroll down once to "Current Targets: Astronomical"
6. Press ENTER to select "Astronomical"

"Parking" the Autostar

If you have performed a good alignment but need to stop observing for a short or long period of time and want to avoid having to realign, just "park" it. Park the scope using the "Park" mode from the Autostar "Utilities" menu before you turn it off. The Autostar remembers the alignment and corrects for the new date and time. The next time you power on, the Autostar will ask for the date, time, and daylight savings. Then it will proceed immediately to the object selection menu.

Software

Autostar Usage

Once aligned the Autostar begins tracking the movement of the night sky. To have the Autostar locate and GOTO to an object, you select the type of object you want to view from a menu of objects (solar system, constellation, deep sky, star, satellite, or user object) and then select the specific object you want to view (e.g., Jupiter). For some objects, such as Messier objects, you can enter the number of the object (e.g., for M42 you would enter "42"). Once the object has been found in the Autostar's database, some information about the object is displayed. You can scroll through this information to learn more about the object or you can press the GOTO button to tell the Autostar to move the telescope to the position of the object. Just as with today's professional telescopes, it will begin slewing to the position of the object in the sky and begin tracking it.

There is a lot of information on the Autostar help screens, accessed via the "?" on the Autostar keypad. When first learning to use your Autostar, be certain to look these over.

The Autostar software and database can be easily updated as Meade release new software or as new objects (such as comets or satellites) appear in the sky. To update the software, you will need a Microsoft Windows-based computer (or an Apple Macintosh running Connectix Corporation's VirtualPC) and a serial cable to connect the Autostar to the computer. Meade have a free Autostar Updater application that downloads software, objects, and "guided tours" into the Autostar handcontroller. Additionally, one Autostar can "clone" itself into a second Autostar. This copies everything from the first Autostar to the second. Both Autostars must be of the same model and you must have the cloning cable (included with Meade's Autostar #505 Connector Cable Set). Finally, there are some astronomy charting software applications that can drive the ETX via the Autostar (see the Appendix for more on software).

You can add your own object GOTO information and satellite data (known as "two line orbital elements" or "TLE") and even make your own "Guided Tours". You can enter object and satellite information directly from the Autostar handcontroller or download via the Autostar Updater application on your desktop or laptop computer. To load Guided Tours, you must use the Updater application. The method of downloading data into the Autostar can change with new versions of the Updater application. Be certain to read the documentation supplied with the Updater software to learn how it is done with that version. To enter user data via the handcontroller, do the following.

Fixed Objects in the Sky

1. Select "User Objects" in the Object menu. Select "Add".
2. Enter the name of the object (use the slewing arrow keys to select and enter characters).
3. Enter the right ascension and declination, angular size, and magnitude.

Satellites

1. Select "Satellite" in the Object menu. Select "Add".
2. Enter the name of the satellite (use the slewing arrow keys to select and enter characters).
3. Enter the TLE information, which can be obtained from many sources, such as the Heavens-Above Web Site (http://www.heavens-above.com). "Epoch Day" is calculated by converting the date and time (i.e., 8:27:08 PM, Sunday, February 25, 2001) to decimal format (i.e., number of days from the beginning of the year; 056.852). "Mean Motion" is also known as "Revolutions per Day". (See "Tracking Satellites" in Chapter 6 for more information.)

GOTO-ing a specific RA and Declination

If you know the right ascension and declination of an object, perhaps tonight's coordinates of a new bright comet, you can directly and simply enter these into the Autostar and have it GOTO that position in the sky. To do this:

1. After the Autostar is aligned and tracking, press and hold the MODE key for 2–3 seconds.
2. Scroll the display (if necessary) to have it display RA and DEC.
3. Press the GOTO key and then the ENTER key to edit the entries.

4. Edit the RA value (as you normally would enter numbers into the Autostar) to be that for the object and press ENTER.
5. Edit the DEC value and press ENTER.
6. If your scope does not begin slewing to the object right away, press the GOTO key.

Resetting the Autostar

There may be times when something goes wrong while using the Autostar. Sometimes, the Autostar will report a "motor failure" or "Proc 2" error. Other times, selected objects may not be accurately centered in the eyepiece or even the finderscope, or random slewing might even occur. The first solution to try is to power off, then turn the system back on, and redo the alignment from the start.

If the problem returns, retraining the drives will usually cure it. When retraining, it is worth doing this with the OTA pointed 45 degrees upward. Apparently this angle adds some extra tension onto the gears.

If you continue to have problems, you may want to recalibrate (which is required if you move the Autostar to a different ETX). This will test the drive motors. Select "Recalibrate Motor" from the Setup menu.

If the problems persist, it may be necessary to RESET the Autostar and re-enter your location and retrain the drives. Select "Reset" from the Setup menu.

If you continue to have problems, the Autostar software may have become corrupted and it may be necessary to download it again. You will need the Autostar Updater software and the connection cable for your model of Autostar.

Note: according to reports posted on my ETX Web Site in 2001, whenever a new version of the Autostar software is downloaded, it is strongly recommended that you RESET and RETRAIN following the download. For more on this technique, see my ETX Web Site.

Other

The various ETX models are extremely portable, but there are some things to consider to keep them performing to expectations.

Traveling with the ETX

Moving the ETX from your living room to your backyard is normally a simple thing. Just pick up the ETX and its tripod, if you have one, and go outside. Even the ETX-125EC and Meade Deluxe Field Tripod are lightweight enough that average strength users will probably have no problem. Of course, the smaller the telescope the easier it is to carry. For some users, it may be necessary to unmount the ETX from its tripod and then remount it once outside. I strongly recommend removing any accessories that may be loose in accessory trays or extra eyepiece holders. You do not want them to fall out while moving the telescope. Once you get set up outside, collect all your accessories and star charts, plug in the AC adapter (if using external AC power), align, and you are all set for an observing session. At the end of the session, you bring it all back inside. Some users have made or purchased special permanently mounted piers (some examples are on my ETX Web Site), and they only carry the ETX outside (along with all their accessories, etc.). Since the ETX is not designed to be left outside in the elements, it is always necessary to bring it back in, even if you have a permanent mount for it.

With the smaller ETX models, their portability means you may actually want to take the ETX with you when traveling, whether by car, bike, motorcycle, train, ship, or plane. You probably will not be taking all your accessories on such trips, so select the ones you really plan to use before beginning your packing. You will want a proper hard- or soft-sided case as mentioned in Chapter 1. If you have the ETX-125EC and plan to take it with you when traveling, a hard-sided case is strongly recommended. Meade urge that you use the foam spacers from the original packing to securely hold the OTA against the base. You should also remove all the batteries (do not forget the battery in illuminated finderscopes and reticle eyepieces) to avoid a problem should they slip out of their holders or leak. Lock both axes but not too tightly. If traveling by commercial means, your telescope and accessories may be subject to being X-rayed or hand-inspected. You really do not want your fine optical instrument transported as "checked baggage", so try your best to keep it with you at all times. If you exercise some caution, your ETX will arrive safely along with you at your destination, perhaps offering new sights to you.

Storage

When not traveling with the ETX and when it is not in use, you may be wondering how to store it. For short periods of time, that is, during the day when you expect to be using the telescope on many clear occasions, you can leave the ETX on its tripod in your home. To prevent any dust from getting on the optics, keep the aperture cover on it and cover any eyepiece in the telescope with the cap from the eyepiece case. Alternatively you can use an empty film can placed into an empty eyepiece hole. If using a film can, do not "drill" into the plastic or metal with the setscrew that secures the eyepiece in the holder. This will prevent bits of plastic or metal from falling inside the OTA. If you want to keep the exterior of the ETX looking as nice as the day you brought it home from the store, then some sort of covering is needed. I prefer a lint-free cloth placed over the telescope to a plastic bag. Some plastic bags will actually "out gas" and leave a residue on your telescope (and optics) as well as possibly cause condensation if left in sunlight. When storing the telescope for longer periods of time, remove all batteries. Do not forget the battery in illuminated finderscopes and reticle eyepieces. If you plan to put the telescope away in a closet, garage, or other storage area, put the telescope and accessories back into their original packing (you did keep it, didn't you?). No matter how long or short a period of time you plan to store your telescope, I recommend unlocking both axes to avoid strain on the locking mechanism.

Chapter 3

Observing Techniques

Preparation

If you just got that new ETX (or any telescope), take the time to read through the manual, several times if necessary, and practice setting it up indoors (where you can see what you are doing before trying it in the dark). Get to know the handcontroller (standard or Autostar) and practice pointing the telescope (slewing) with the controller. Learn how to exchange eyepieces if you have more than one. Do not forget to align the finderscope as well as you can; this will be a tremendous help in locating some objects when you go outside. You should do the finderscope alignment with the telescope outside and use a fixed object several miles away (if possible). Put the object in the center of a moderately high-power eyepiece, lock both axes, and then begin fiddling with the adjustment screws on the finderscope mounting bracket. I find it easiest to loosen just the front or rear ones, leaving the screws at the other end tight. Then begin tightening the screws until the cross-hair is nearly on top of the object. I say "nearly" because you really do not want it exactly on top; the thick cross-hair can actually cover a star or other object you are trying to center. In this case, "close" is good enough. As part of your initial learning, you should begin to get at least some familiarity with the night sky and some commonly used astronomical terms. Much of this is contained in the telescope's manual, but there is a lot more that you can learn. See the Appendix for some sources of information that can be valuable to you. Take your time; do not try to absorb everything in one sitting. Spend as many days as necessary to learn your telescope. If it is cloudy when you first get your telescope, do not fret; unpack it and learn it. Use this time to your advantage. If you spend this time upfront, you will feel more comfortable and less frustrated when using it in the dark.

OK, so now you know how to use your telescope and you at least know which way is north. Are you ready to go outside and observe? Well, maybe. If you know the Moon is up and you want to see it "close up", go observe it. But what if Venus was also visible, and you started with the Moon? By the time you got around to looking at Venus in the west, it might be too low to be easily seen or best viewed. Had you looked at Venus first, you would have been rewarded with a view of a beautiful gleaming white crescent-shaped planet. Oops. Maybe you should have planned things out better.

Planning for a night's (or morning's) observing session means: knowing what you want to observe, in what order to look at them, where they are in the sky, and, to some extent, what you might expect to see and what eyepiece you plan to use with each object. If you are new to astronomy, start with the brighter, easier-to-find objects to gain experience in using your telescope to locate objects (some techniques are described later in this chapter). You will also need to gain experience in observational techniques (also discussed later in this chapter). Some things to consider when planning a session are the Moon's phase, the weather, and the position of objects in the sky in relation to structures on the ground (lights, heat sources, or obstacles). If you plan to observe faint objects, you need a dark sky. If the Moon could be a factor, plan to observe fainter objects after it has set or before it rises. Weather can be a factor. If nightly fog is prevalent in your observing location, you will want to plan to observe before it rolls in. The same applies if there are clouds in the forecast; get that observing done before it gets cloudy. One often over-looked aspect of planning an observation session is objects on the ground. Trying to view Mars through a building is pretty difficult. Turbulence due to heat rising from a roof or parking lot can ruin your view almost as badly. If you are trying to view some faint object to the southeast but there is a streetlight in that direction, you might find it difficult to locate the object. If your neighbor turns off his yard light at 11pm, you will find observing after that more rewarding. So plan your sessions according to your environment.

If you have an Autostar, you can use it to GOTO objects for you. You can either select the objects yourself or use the "tour" mode to have the Autostar select them for you. If you do not have an Autostar or you want to learn more about the night sky, you will want to check a star chart for what is visible at the times you want to observe or when an object you want to see will be visible.

OK, now you know what and when you will be observing. Are you ready to go outside? No, not yet. You need to determine which mounting mode is to be used for what you plan to do, assuming your equipment provides you with a choice.

Polar vs Alt/Az

The ETX can be used in either polar mode or alt/az mode, with or without an Autostar. Both modes have their pluses and minuses, and sometimes one mode is required.

In polar mode, also known as "equatorial mount", the ETX base vertical rotational axis is tilted to be parallel to the Earth's rotational axis. In this mode, the ETX azimuth drive (known as the "right ascension" drive in this mode) runs constantly at the same rate as the Earth rotates. Assuming a perfect alignment, fixed astronomical objects (e.g., stars, galaxies, and nebulae) will stay in the same position in the eyepiece for hours. However, in practice, objects may stay centered for only minutes or as long as an hour. This is because most users do not go to the trouble to precisely point the telescope to True North (not always the same direction as your magnetic compass shows as north) or to precisely set their latitude. If you take the time at the start of your observing session to make a reasonably accurate set-up (as described previously in Chapter 2), you will be rewarded with excellent tracking throughout your session (assuming you do not bump your tripod and have to realign it).

In altitude/azimuth (or alt/az for short) mode, the ETX base vertical rotational axis is pointed straight up. In this mode, the altitude and azimuth drives need to be used together to keep celestial objects centered in the eyepiece. If the two drives are not used together, objects will drift out of view in just a few seconds, depending upon the magnification being used. As mentioned earlier in this book, the Autostar can properly track using both drives when the ETX is mounted in this mode.

Why use one mode over the other? As it turns out, there are many reasons, from the equipment you have on hand to how you plan to use the ETX. If all you have is the basic ETX without a tripod or the tabletop legs, you will be using the ETX in alt/az mode, typically with the ETX sitting on a flat surface, such as a backyard table or even a concrete patio. If you have a camera tripod, it might not be stable enough to support the ETX when it is tilted to match your latitude; so you should use alt/az mode, which keeps the center of gravity of the ETX over the central portion of the tripod. Then there is how you want to use the ETX; for terrestrial viewing or photography, or celestial viewing or photography.

When looking at objects on Earth (known as "terrestrial" viewing), you do not have to worry about compensating for the Earth's rotation; so alt/az mode with its simpler mounting requirements is the way to go. You can use a handcontroller (the standard one or the Autostar) or manually slew the ETX field-of-view left and right, up and down, to locate objects. For astronomical purposes, things get more complicated. If you have an Autostar-capable telescope and you have an Autostar, you can use either polar or alt/az mode if your tripod can handle either mounting method. Most ETX users with the Autostar prefer alt/az mode for its ease of set-up and stability. Other users prefer to use polar mounting even though they use the Autostar since only one drive motor has to run to track, thereby decreasing motor vibration of the image being viewed. If you do not have the Autostar or you have the original ETX-90RA, then you need to use polar mode if you want the ETX to track objects automatically.

If you plan to try long duration astrophotography with your ETX/Autostar system, you will need to use polar mode to avoid the effect known as "field rotation", which would result in undesirable smearing of images within the field-of-view on the film or CCD. Field rotation occurs when the mounting is alt/az and the telescope tracks astronomical objects by moving in both altitude and azimuth. Using polar mode avoids this effect, since tracking is in one direction only (right ascension) and images do not rotate in the eyepiece. For visual observing, field rotation is not a problem.

There is another consideration in determining which mounting mode to use and that is the ETX eyepiece and finderscope locations. With the design of the ETX, it is possible that some objects cannot be observed in some ETX orientations. For example, at some latitudes, you will find it impossible to view some objects above the southern horizon (in the northern hemisphere) when the ETX is mounted in polar mode. This is because the ETX base is tilted upwards, and the telescope tube hits the base before it can move completely to the southern horizon. As you move to higher latitudes, this becomes less of a problem. This is not a problem in alt/az mode where the base is horizontal, but you may find that using the finderscope is difficult when the telescope is pointed at the zenith (straight up). If you have a mounting choice, you may want to consider object location versus mounting mode for ease of viewing.

Locating Objects

Now that you have your observing plan for the night established and your telescope is set up, how do you find that object with your telescope? If you have an Autostar, you may think you can let it do all the work for you; but you might be wrong. It might put that "faint fuzzy blob" right in the middle of your 4 mm eyepiece, but you might not be able to see it. Why not? I will come to that eventually, but for those without an Autostar, I should begin at the beginning.

Many objects in the sky are easily located and observed. For example, the Moon and the brighter planets can usually be located by the newest of amateur astronomers. But what about a faint nebula or galaxy, or even a specific double star? How do you locate them with your telescope without a GOTO computer? First you need a chart showing the object's position in relation to other bright and faint objects around it or you need to know its right ascension and declination coordinates, and sometimes both a chart and the coordinates are useful. Next you go "star hopping", described in the next paragraph.

Looking at the star chart, pick a bright naked eye star that is reasonably close (that is, within a few degrees) of the object you want to locate. Next find the object you want on the star chart. Note the "path" between the first marker object and your target object. Pick additional "marker stars" along this path, noting anything special

about them that makes them easy to identify (i.e.,
brightness, patterns of nearby stars). As you get close to
your desired object, you may need a more detailed
chart to pinpoint its position as precisely as possible.
With your chart in hand (or committed to memory), go
to your telescope. Use a low-power and/or wide-field
eyepiece. Locate your first marker in the sky, then point
your telescope at its location and center it in the
finderscope. If your finderscope is properly aligned, you
should see this object in your eyepiece. (If you do not
see the object in the eyepiece, now would be a good
time to align the finderscope.) Now slowly slew the
telescope to the second marker on the path to your
desired object. Keep in mind the image reversal when
viewing through an eyepiece (if appropriate). Ideally
each next marker you selected should be in the eyepiece
field-of-view before the previous one goes out of view.
This lets you follow a well-defined path without getting
lost. Watch for any special items that you noted when
you selected the markers on the star chart to verify you
have found the right marker. As you "star hop" from
one marker to the next, you will eventually come to the
location where the object should be visible in your
eyepiece. I say "should be" because it might be there
but you might not be able to see it for reasons explained
in the next section. But if everything worked out, the
highly prized object you sought will be there for you to
appreciate. If it is not and you know that you should be
able to see it, start over with your first marker object.
Even experienced "star hoppers" can miss a star
marker or mis-identify one. This can throw your path
off just enough to lead you astray. As you gain
experience with this star-hopping technique, you will
learn to pick better marker stars and be able to identify
them more quickly in the eyepiece. Do not try for the
faintest possible objects until you have some experience
star hopping to brighter objects.

If you know an object's right ascension and
declination, you can slew the telescope to its location
using the RA and DEC setting circles. This requires that
the setting circles be accurately calibrated. It also
requires that your telescope be mounted in polar mode,
whereas you can use the star-hopping method in either
mounting mode. You will need to know not only the
desired object's RA and DEC but the coordinates of a
bright star in the same portion of the sky (i.e., within a
couple of hours of RA and 30 degrees of declination).
With the right ascension drive engaged to track the

sky's movement, put that nearby bright star in the field-of-view of a moderately high-power eyepiece. Next, verify that the declination setting circle shows the same value as this object's declination. If the setting circle shows something different, do not worry about it now. Just remember the difference and carry that along as you point to other objects. Later (that is, tomorrow) you can adjust the position of the setting circle as described in Chapter 2. When you finish with the adjustment, verify that the object is still in the eyepiece. Next you must adjust the right ascension setting circle to match the object's right ascension coordinate. The RA setting circle on the ETX (and most small telescopes) is moveable. On the ETX it slides along in its track on the base. Using your finger or, if it is stuck, the eraser end of a pencil, slide the tape with the numbers on it until the object's right ascension value on the setting circle is beneath the proper marker on the base of the telescope. Verify again that the object is still in the eyepiece. If not, repeat these steps. Once you have both setting circles calibrated, you slew the telescope until the positions of the setting circles match the RA and declination of your target object. If you accurately calibrated the setting circles, the object should be in the eyepiece. I say "should be" for the same reasons as when star hopping, but there can be some more reasons that the object might not be in the eyepiece when using this technique. If the polar alignment is not accurate, the direction in which the scope is moved may not be the actual direction from the first object to the target object. Accurate polar alignment is normally required for best results. A common problem is that the accuracy of this method depends upon the accuracy of the setting circle calibration.

Since the setting circles on the ETX are fairly small, it is difficult to get much accuracy. The RA scale has division markers every 5 minutes of right ascension and the DEC scale every degree. It is also possible that the RA setting circle slipped during the slew to the object. And finally, as the ETX RA setting circle has two sets of numbers, one increasing as you go around the tape and the other decreasing, it is easy to use the wrong set. The numbers on the bottom are for use in the northern hemisphere and the ones on the top are for the southern hemisphere. But if you cannot keep that straight, just remember that RA increases as you go eastward. This technique can work and is what astronomers have used for years (prior to computer controlled telescopes). In fact, it is what I used to locate Venus during the daytime for the photo in Chapter 4. Using proper protection, I calibrated the setting circles for the Sun's position and then slewed to the coordinates for Venus. I did have to slew around a little bit to actually locate Venus, but I knew that I was in its neighborhood. So, if your object does not appear in the eyepiece, slew back and forth and up and down in small increments. Slew slowly, stop frequently, and look around the field-of-view.

In many cases, you will actually use a combination of the setting circles and star hopping to locate target objects. The setting circles will get you close and then you use star hopping to refine the pointing.

Things To Be Aware Of

You or the Autostar have done everything correctly and you still cannot see the object even though you know it absolutely must be in the eyepiece. What could be wrong now? Actually, there are many things that could be keeping you from seeing the object, some of which you have control over and some you do not.

First of all, is the object even expected to be visible in your telescope? If its magnitude is fainter than the limiting magnitude of your telescope, you will probably not be able to see it except under unique circumstances. Also, the higher the magnification you are using the more likely those faint objects will be invisible; try a low or moderate power eyepiece. Perhaps an eyepiece with a wider field-of-view will let you better identify the star patterns around the object, which in turn can make the object stand out.

Vibrations in the telescope can affect the ability of the eye to see a faint object. These vibrations can come from focusing, slewing, the motor drives, the wind, or even heavy trucks passing by your location. A sturdy mount can reduce the possibility of vibrations ruining your view. Following a slew, let any vibrations dampen out before giving up on being able to see a faint object. If possible be certain you have focused the eyepiece on a different object in the same area of the sky (to avoid focus shifting as the telescope orientation is changed), so that you do not have to refocus while looking for a faint object.

Sky conditions can also affect what can be seen. Obviously, clouds or fog can effectively hide an object from view. If you cannot see any stars in the eyepiece, look up! I cannot count the number of times when I have not been able to see an object only to discover that clouds or fog had moved into that portion of the sky. Another condition that can affect your viewing is atmospheric turbulence, commonly called "seeing". Whether the turbulence is due to local heat sources (roofs or parking lots) or motion higher up in the atmosphere, if brighter objects appear to be "boiling" or alternately blurring and sharpening when viewed through an eyepiece, then you can expect to have difficulties seeing some objects, faint or not. This turbulence can really detract from viewing, and when severe you should probably give up for the night and go read a book on astronomy. One other often overlooked culprit of poor seeing is dew forming on the optics. Even if you have a dew shield on your telescope, check all the optics if you suspect that the view has deteriorated during a session. You may find that one or more of the optical surfaces have clouded up with moisture, even from your breath. Sky brightness can also affect what can be seen. If the Moon is bright, faint objects will likely not be visible. If there are a lot of streetlights or other lighting in your neighborhood, you should expect difficulties locating faint objects. If you are attempting to locate really faint objects, a really dark sky is of paramount importance.

Air turbulence in the atmosphere can affect the views but so can air turbulence inside the telescope tube. If you have just brought the telescope outside from a warm house, for several minutes to as long as an hour, air currents inside the tube can cause a lot of visible distortion in viewed objects. Most times many observers never worry about this, either because they do not realize it is occurring, or because they are in a hurry, or because it is not so severe as to be bothersome. For the best views, you should let the telescope reach what is called "thermal equilibrium" with the surrounding air. You can accelerate this process by moving the telescope outside well before planning to observe and leaving the eyepiece hole open to allow better air circulation.

Finally, there is one optical instrument whose condition is extremely important for good viewing: your eyes. If you want to see faint objects, you absolutely must let your eyes become "dark adapted" and continue to avoid any lights except that from a red-lensed flashlight. This can take from 15 minutes to an hour depending upon the condition of your eyes and the lighting they are subjected to. Attempting to see a faint object by staring at it is likely to be futile. In low light situations, the central portion of the eye's retina is not as sensitive as other areas. Using "averted vision", where you look slightly to the side of where the object should be, may let you see an object that would otherwise be invisible to you. If you look, then wait, and then look again you can let your brain "build up" the image allowing you to see more details. Using these techniques you may be surprised at the details you will see in some faint objects.

Recording What You See

OK, so now you have located the object and you are thrilled at what you see. Is that sufficient? For many amateur astronomers, it is completely satisfying and sufficient. They want to see as many objects as they can and just know that they saw them. For others, locating the object is just the beginning of the enjoyment of astronomy. If you are in this latter category, then read on.

If you decide that you want to record your observations, you need some technique to do this. You can develop a technique that works for you or use all or some of the techniques described here. I make no claim that these techniques are any better or worse than others.

The first thing users normally start recording are some facts about the observation. They keep the star-hopping chart they prepared and add some information to it. Typically this is the date and time that the observation was made, the location (home, longitude/latitude, etc.) that the observation was made from, and the equipment (telescope model, eyepiece, filters, etc.) used to make the observation. It is helpful to record the sky conditions (including the Moon's phase and nearness to the object being observed) at the time of the observation. The more detailed you make this observation log the more useful it will be when trying to view this same or similar object in the future.

A nice observing log layout and database system to store your observations for later retrieval has been developed by David Green (http://www.davidpaulgreen.com/) and is shown in Figure 3.1 (used with permission). His "The Simple Observing Log" for Macintosh and Windows uses the FileMaker Pro database to store your observations. If you do not have FileMaker Pro, David also provides a standalone version.

Figure 3.1. The Simple Observing Log.

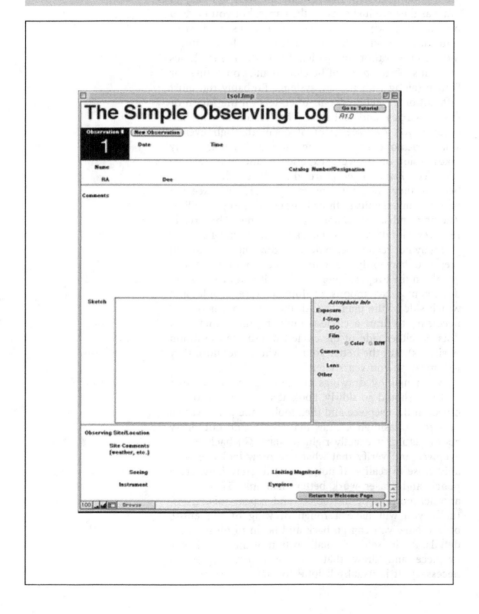

The next step up from logging the observation is to actually make a drawing of what you see through the eyepiece. You do not have to be an artist to do this, but it does take some practice to do it well. Start by drawing brighter, less detailed objects, for example the shape of the phase of Venus. After you become good at drawing reasonable likenesses of simpler objects, you can progress to drawing more detailed images. Star patterns are a good next step. Individually the stars are easy to draw as they are just dots on the paper, but you have to put them down in the proper patterns and sizes (indicating magnitude or brightness). This requires more concentration and patience than doing the phases of Venus. Next up might be cloud bands on Jupiter or Saturn (along with its ring system). Probably the most difficult of all will be details of craters on the Moon or faint galaxies and nebulae. The Moon is difficult, because you can see a lot of detail through even a small telescope and drawing all those details really takes a lot of drawing experience and time at the eyepiece. Faint objects are difficult to draw, well, because they are faint. You will probably not see too many, if any, details in these faint objects, so you will be drawing indistinct shapes and shading. This really requires a lot of patience and skill to do well. Of course, you may not really need a detailed drawing of what you see; you may only want an indication of what was visible in the eyepiece. For example, if you just want to document what moons of Jupiter were visible and which side of the planet's disk they were on, it is only necessary to draw a circle (simulating the planet) and dots on either side of the circle (to indicate the moon positions). But the better your drawings, the more they will mean to you years later.

When making drawings at the eyepiece, remember that you should go slowly. Look at some portion of the object in the eyepiece and then look at the paper (use a red-lensed flashlight). Draw that portion; do not worry about getting it exactly right to start. Go back to the eyepiece and verify that what you drew looks reasonably close to reality. If not, correct it. Here is where a pencil and eraser work better than ink. Then go to another portion of the object and repeat this. Eventually you will have a rough drawing of the entire object. Now you can go back and begin to fill in some details. Again, select a small portion of the view in the eyepiece and draw that on the paper. Repeat as necessary. It is usually helpful to draw a surrounding circle to indicate the edge of the eyepiece field-of-view. I find it constricting to have this edge drawn before actually drawing the object, so I add it at the end. Others prefer to have the edge drawn first to keep them at the proper scale when drawing the object. Try both and select the method that works for you.

If you find that logging and drawing distract you from enjoying your time at the telescope, skip them. Astronomy is fun and its rewards are varied. Do what you enjoy and do not get hung up on things that may be unimportant to you.

Of course, the ultimate in recording observations is photography. Unfortunately, the nature of astrophotography is such that the camera usually does not record objects as you see them with your eye. Objects like the Moon may appear similar in photographs to what your eye sees, but fainter and smaller objects may appear worse or better in photographs. Astrophotography is discussed in more detail in Chapter 5.

Chapter 4
Objects

In this chapter, I will discuss 100 of the best objects for the ETX. Photographs and drawings for some object types are used to simulate what the eye sees. As discussed in the previous chapter on observing techniques, when using a telescope to observe distant objects, many variables can affect the quality of the viewed image. Even fatigue can affect what the eye/brain combination can perceive. So your experiences may differ slightly from what I present here. Conditions may not have always been perfect when the photos or drawings were made; but they are representative of what you can expect when observing from a typical, moderately dark or near urban locale sky. Your results may be better or worse. You may or may not be able to see some of the objects described. You may have other "best objects" that you enjoy looking at with your telescope. The selection here is to provide the new amateur astronomer with some objects that can be seen with a small telescope like the ETX-90. The results when using a telescope with a different aperture size or focal length will likely be different from what is noted here. Some of these objects may be easy to locate and see, even under less than ideal observing conditions. Some may be a challenge, requiring the best observing conditions and some of the techniques discussed in Chapter 3. It is good to know that someone else has observed a particularly challenging object with the same telescope you own.

I will start with descriptions of the easier objects and then proceed to more difficult ones. Until you gain some observational experience, you should also start with the easier objects and then proceed to the more difficult ones. Remember that gaining this experience is part of the fun of amateur astronomy.

One comment about the photographs and drawings: they were made using my ETX-90RA. The eyepiece used for each is noted, but in general I tried to use only the standard Meade 26 mm Super Plossl eyepiece that comes with the ETX. This eyepiece provides a visual magnification of 48× on the ETX-90. Sometimes more magnification was needed and so the Meade 9.7 mm eyepiece (128×) was used.

The Moon

First up is an object that is easily viewed and changes its appearance daily: the Moon. The Moon makes an excellent first object when beginning to learn amateur astronomy or using a telescope for the first time. You can view it from almost any location, under a variety of conditions. As the Moon revolves around the Earth, its "phase" or the illuminated portion seen from the Earth, changes. It proceeds from a New Moon (totally dark and usually invisible) to a thin crescent where only a small sliver of the Moon's surface is illuminated by sunlight (and the rest illuminated by sunlight reflected from the Earth, known as "Earthshine") to First Quarter where half the visible surface is illuminated by sunlight to Full Moon; and then the process reverses through Last Quarter Moon, a thin crescent Moon, and finally New Moon. This process repeats on an approximately monthly basis (28 days). At each step along the way, new sights become visible and others change. In more brightly lit areas on the Moon, you will see "rays," long white lines and arcs of material ejected from impacts on the lunar surface. Finally there are the lunar "plains" (also called "seas") that are large areas of generally flat, light and dark lava flows. See Figure 4.1 for an example of what you can see at Full Moon. During the partial phases where the shadows are the longest, details of lunar craters and mountains are most evident at the terminator (the dividing line between the illuminated portion and the dark portion). Using higher magnifications is required to see these details along the terminator. See Figure 4.2 for a typical view along the terminator. Many users learn all the features on the Moon by name, using a lunar chart (a simple version of

Figure 4.1. Full Moon as seen through the 26 mm eyepiece.

which is shown in Figure 4.3). The ETX is an excellent telescope for observing the Moon. Some users find it useful to add filters to the eyepiece (polarizing or neutral density) to decrease the brightness and glare when viewing the Full (or nearly so) Moon using the 26 mm eyepiece.

Lunar eclipses, where the moon moves into the Earth's shadow, are real treats. Each eclipse is different, depending as they do upon the orientation of the Moon's path through the shadow and the condition of

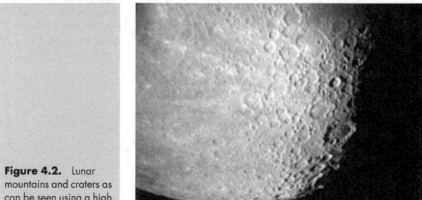

Figure 4.2. Lunar mountains and craters as can be seen using a high magnification eyepiece.

Figure 4.3. Simple Moon chart. Reproduced from Cook (ed.) The Hatfield Photographic Lunar Atlas, Springer.

the Earth's atmosphere. Some eclipses are very dark. At other times the Moon may take on a distinct reddish cast. Some users enjoy timing the entry and exit of lunar features into and out of the shadow. Others just enjoy the show. The ETX makes a good telescope for both types of observers. Photographing lunar eclipses is also fun and nearly any camera, 35 mm film, digital, or even a video camera can be used to capture the eclipse.

Occultations of the planets and stars are fascinating to watch. An occultation occurs when the Moon passes in front of an object, hiding that object from view. An object can be totally hidden or can "graze" along the Moon's limb, blinking as it goes behind mountaintops and reappears. Grazes of bright objects are rare, so catch them when you can. Normal occultations have the object getting closer to the Moon (as the Moon moves

Figure 4.4. Lunar eclipse photograph.

eastward in its orbit around the Earth), then disappearing, either slowly behind the Moon (for a large object like a planet) or rapidly (for a star), and then eventually reappearing on the other side of the Moon. Depending upon the conditions of the event, the object may disappear or reappear on the illuminated side of the Moon or along the dark, unilluminated side. Viewing a disappearance is easy, just locate the object prior to it reaching the Moon's limb and then follow it until it disappears. Reappearances are more challenging to watch since you have to know exactly where along the Moon's limb to look. For this you will need a finder chart from some source (magazine, Web Site, astronomy software, etc.) showing where along the Moon's limb the object will reappear. Disappearances and reappearances along the dark side of the Moon's disk are especially fun to watch, since the object seemingly disappears into "thin air" or magically reappears "out of nothing".

Figure 4.5a shows two digital camera photographs taken of the September 1997 occultation of Saturn. This occultation occurred shortly before sunrise. The two photographs show the Moon's illuminated side approaching Saturn and then beginning to cover the planet.

Figures 4.5b to 4.5f show the July 2001 occultation of Venus. The Venus occultation was particularly challenging since it occurred shortly before noon at my location. The Moon was a thin waning crescent and Venus was in a gibbous phase. The Venus occultation

Figure 4.5a.
Occultation of Saturn.

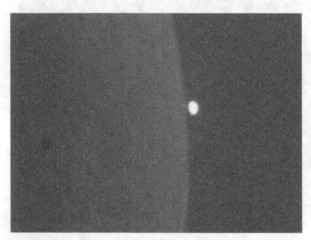

Figure 4.5b.
Occultation of Venus.

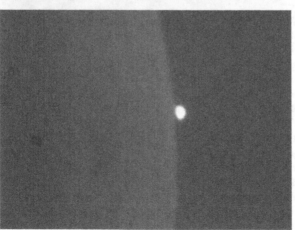

Figure 4.5c.
Occultation of Venus.

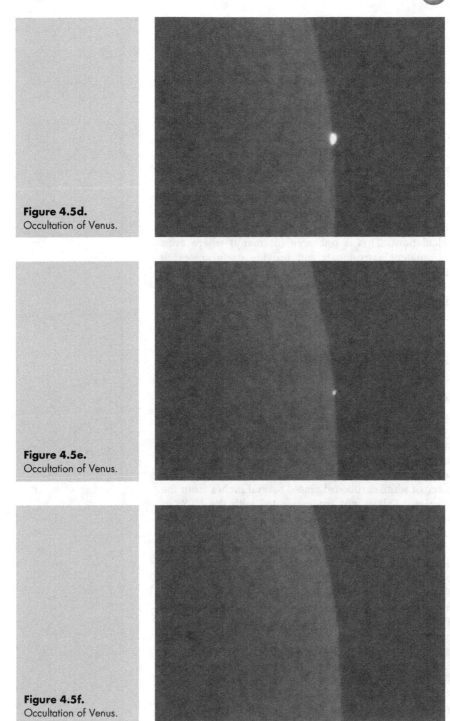

Figure 4.5d.
Occultation of Venus.

Figure 4.5e.
Occultation of Venus.

Figure 4.5f.
Occultation of Venus.

photographs are still images from a video captured by the Sonfest Promotions, Inc., (www.sac-imaging.com) SAC IVb CCD imager using an ETX-125EC.

These photographs provide a good idea of what you can see during a planetary occultation using a magnification of 128×.

Taking good photographs of occultations is difficult, especially when the Moon's surface at the disappearance or reappearance location is illuminated. The difference in image brightness between the two objects usually causes either over-exposure or under-exposure of one of the objects.

Some amateur astronomers make precise timings of occultations and send these to professionals, who use them to make very accurate orbit and position calculations. This is one area (of many) where even an amateur astronomer can provide useful research data. More on occultations and predictions are available at the International Occultation Timing Association Web Site (http://www.lunar-occultations.com/iota/iotandx.htm).

The Sun

With the proper protection, both for your eyes and your telescope, viewing the nearest star, our Sun, can be rewarding. However, failure to heed some precautions can result in serious eye damage, including blindness, and may even damage your telescope. With some telescopes you can project the Sun's bright image onto a piece of white cardboard placed several inches from the eyepiece. This is **not** recommended with the ETX, as you run the risk of the intense heat melting parts of the OTA. The preferred method to view the Sun with the ETX is by using a quality solar filter from a reputable dealer (see the Filters discussion in Chapter 2). I purchased a Solar II Type 2 Plus solar filter from Thousand Oaks Optical in California. Since I wanted it for both visual and photographic use, I went with the Type 2 Plus, which has a transmission factor of 1/1000 of 1% (ND-5). The filter is glass, coated (reflects like a mirror), felt-lined for a slip-on but secure fit on the ETX, and provides a nice yellow-orange view of the Sun. The filter covers the entire 90 mm aperture of the ETX-90 (other sizes are available for other telescopes). You should also make a cover for the finderscope to prevent its accidental use. Always inspect the solar filter prior to each use to ensure its coating is undamaged. Do not use a damaged filter!

With the solar filter and finderscope cover in place, getting the Sun into the eyepiece field-of-view (FOV) can be challenging at first. I use the shadow of the ETX. When various protrusions (like the eyepiece or finderscope) appear to have their shadows minimized, I check the eyepiece. With practice you can get at least a portion of the Sun in the 26 mm eyepiece FOV. Once you have the Sun in the eyepiece, you may see a view like that in Figure 4.6, which was taken on 13 May 2000 with a digital camera. Several sunspots are easily seen across the entire solar disk (the fainter splotches are artifacts from the exposure).

Figure 4.6. The Sun with many sunspots.

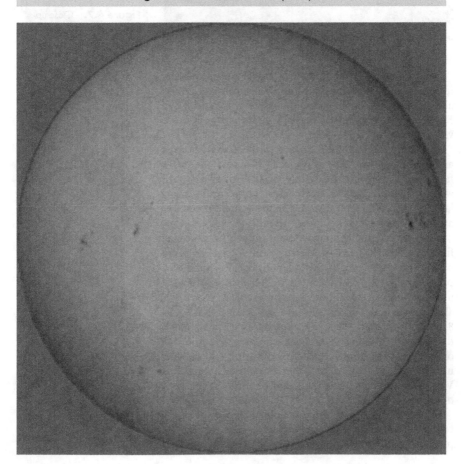

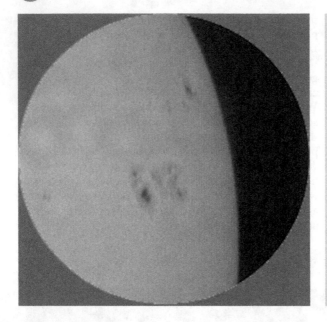

Figure 4.7. Close-up of a sunspot grouping.

You can increase the magnification to a point. As the image getting past the filter is quite dim, you will not be able to go much beyond the 9.7 mm eyepiece with the ETX-90. But even at that magnification, you will see a view similar to Figure 4.7.

Watching the changing surface of the Sun over days and months is fun. As the spots drift across the Sun's face, you are seeing the effect of the solar rotation, which is about 25–35 days (depending upon solar latitude). Keep a log and/or take photographs if you are a serious Sun watcher.

Another exciting thing you can do if you have a solar filter is watch transits of Mercury and Venus as these planets pass in front of the solar disk. I observed and photographed the 15 November 1999 transit of Mercury using my ETX-90RA (photos are on my ETX Web Site). The next couple of transits are on 7 May 2003 and 8 November 2006. Mercury appears as a small black disk against the solar surface and moves across the Sun over a period of hours (the actual duration will depend upon Mercury's path across the solar disk). Transits of Venus are rarer and so should not be missed. The next ones are on 8 June 2004 and 5/6 June 2012.

Solar eclipses are exciting events that do not really require a telescope. You can use your ETX during the partial phases as the Moon passes in front of the Sun (but do not forget to use a solar filter). As totality approaches you should stop looking through the telescope and enjoy the change in your surroundings. During totality you can use the ETX to view comets and planets near the Sun, but you will likely spend the whole time just looking at the view of the solar corona and photographing it (use a telephoto lens, not the ETX). You can use the ETX to take photos of prominences (massive "explosions" on the Sun's surface) that may be visible during totality along the edge of the Sun's disk. During totality you will not use the solar filter, but be prepared to swing the ETX away from the Sun at the end of totality. Total solar eclipses are rather rare, but partial eclipses are more frequent. One recent partial eclipse occurred on Christmas Day 2000 (see Figure 4.8). From my residence in southern California, only 15% of the Sun's diameter was covered at maximum eclipse but it was nevertheless an exciting event to watch and photograph.

To see or photograph prominences when there is no solar eclipse will require a hydrogen-alpha filter. These are usually expensive (typically more than the cost of the ETX), but if you are really into Sun watching you may want to invest in one. In 2001, Coronado released a new low-cost (less than $1000) H-A filter for use on many telescopes, including the various ETX models. As

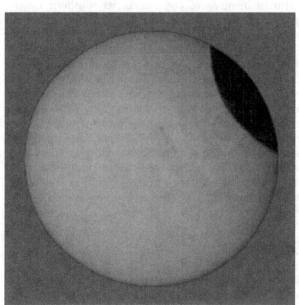

Figure 4.8. Partial solar eclipse.

this is being written, I have had the opportunity to view prominences and many other details on the solar surface using one of these new filters. Watch my ETX Web Site for more on this filter.

The Planets

The planets make excellent targets to expand your telescope use skills. Typically there are some planets that are easily visible for about six months during the year. At other times, these same planets are lost in the Sun's glare. The brighter planets like Jupiter, Saturn, and Venus are beautiful to look at and present their own unique views.

Jupiter

Jupiter is the largest planet in our solar system with about 10 times the Earth's diameter. When looking at it with the 26 mm eyepiece or one with higher magnification, the first thing that most users will normally see are cloud bands in the Jovian atmosphere. These bands, running parallel to Jupiter's equator, vary in size and color but at least two (one in Jupiter's northern hemisphere and one in its southern hemisphere) are always visible, and sometimes you may see more. Next, depending upon their orbital positions, one, two, three, or all four of the brighter Jovian moons (Ganymede, Europa, Callisto, and Io) may be seen. These moons appear as small star-like objects. If you watch over several hours, you may detect the rotation of the planet (about 10 hours for one "day") by changes in the cloud bands and see movement of some of the moons. The famous "Great Red Spot" has faded in recent years and so is more like a "pale pink spot" and is extremely difficult to see without additional aids (filters). Jupiter is difficult to photograph as it appears to the eye so a drawing (Figure 4.9) is provided.

Saturn

Saturn and its ring is an impressive sight in almost any telescope. Saturn is the second largest planet in our

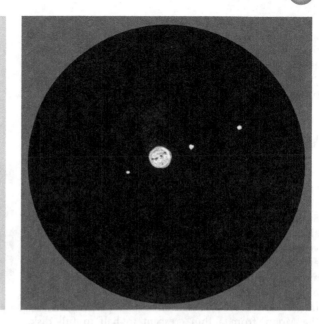

Figure 4.9. Jupiter and three moons in the 26 mm eyepiece.

solar system. With good seeing and excellent optics you can detect some cloud bands on the planet, a dark line called "Cassini's Division" running around the center of the ring "surface", and perhaps the ring's shadow on the planet and the planet's shadow on the ring. As

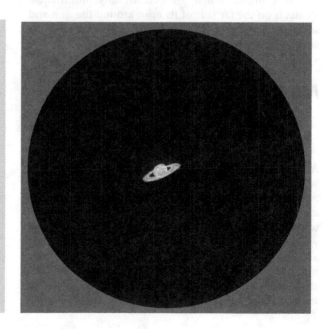

Figure 4.10. Saturn and its rings in the 26 mm eyepiece.

Saturn moves along in its nearly 30-year long orbit, you will notice that the angle of the rings changes as viewed from the Earth. Sometimes the rings will be edge-on, essentially invisible, and at other times present a very "wide-open" appearance. You may also see Saturn's brightest moon Triton.

Venus

Popularly known as the "Morning Star" or "Evening Star", Venus is not a star but a planet almost the same size as the Earth. But at times in its orbit around the Sun, it does appear to be the brightest "star" in the sky shortly before sunrise or after sunset. In fact, during very favorable conditions, the light from Venus can even project shadows. Its close orbit around the Sun and its planetwide cloud cover make it a very bright object as it reflects a lot of sunlight. Since its orbit is inside Earth's orbit, Venus manifests phases just like the Moon, from a thin crescent to half to full disk. Figure 4.11 shows a crescent Venus. This digital camera photograph was actually taken during the afternoon at a magnification of 128×.

Figure 4.11. Venus.

Venus is at its best when closer to the Earth. It is then that it is largest and brightest even though in a less than half disk phase. When the disk is fully illuminated, Venus is on the far side of its orbit around the Sun and so is farthest from the Earth and therefore is much smaller in apparent size and is fainter.

Mars

Mars is best seen when in "opposition" (when its orbit around the Sun places it opposite the Earth) and is therefore closest. Some oppositions are better than others. At the best oppositions, Mars can appear very nice in a 90 mm telescope whereas at most oppositions, it will present a disappointing view in a small telescope. Usually, the most noticeable thing about Mars will be its color: it will appear to be a small ruddy red disk. At its best you will be able to see some dark areas on the planet's surface and one of the polar ice caps. If you observe Mars over many months, you can actually see the effects of its changing seasons, which are just like the seasons (spring, summer, fall, and winter) on the Earth. Dark areas will expand or contract, and you may be able to witness major dust storms that occur on Mars, blocking out the view of the dark areas. Mars is a challenging object, especially in a small telescope, but it can also be a rewarding one with many treats for the dedicated observer.

Mercury

A challenge to see even at its best, Mercury is a planet that many have never seen. Due to its close orbit around the Sun, it never gets far from the Sun and so appears very low in the western evening or eastern morning sky. And because it is so low, atmospheric turbulence will disturb the view and make Mercury blurry at times. It is also a small planet, physically about the size of our Moon. Like Venus, since its orbit is inside the Earth's orbit, Mercury exhibits phases. It can be bright or dim depending on its phase. Occasionally, Mercury has a "favorable elongation" when it is far from the Sun in the sky and far from the horizon. Other times are less favorable because the plane of its orbit keeps Mercury very low and near the horizon.

Such a favorable elongation occurred in June 2000. Figure 4.12 shows the "half moon" phase that Mercury displayed at that time.

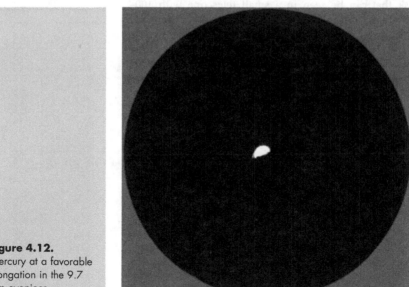

Figure 4.12.
Mercury at a favorable elongation in the 9.7 mm eyepiece.

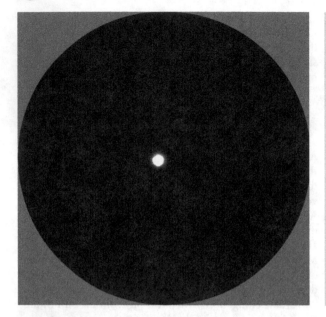

Figure 4.13. Uranus in the 9.7 mm eyepiece.

Uranus

Another planet that few have seen, Uranus is a fun object to go after. It is about 10 times larger than Mercury, but it is about 1.7 billion miles further from the Sun than Mercury. Uranus shows a small bluish disk that can be seen in a small telescope with 100–150× magnification. The best time to view Uranus is when it is at opposition. At that time, it appears high in the sky around midnight, so observing conditions are ideal. Figure 4.13 provides you with an idea of what Uranus can look like in an ETX-90 at 128×.

Neptune

Normally, the eighth planet out from the Sun in our solar system, Neptune is a real challenge for the small telescope observer. It is small and faint, being about 1 billion miles further out from the Sun than Uranus and a little smaller in diameter than Uranus. To some it will look greenish, to others bluish, and to many it will have no color at all. At its best, and with good seeing allowing magnification in excess of 100×, you can see a small disk. Figure 4.14 is a drawing of Neptune made at the 9.7 mm eyepiece on the ETX-90RA (128×).

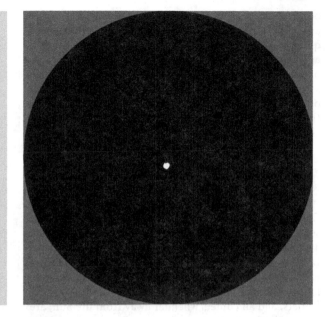

Figure 4.14.
Neptune in the 9.7 mm
eyepiece.

Pluto

This last planet of our solar system is not just a
challenging object in a small telescope; it is an
impossible object to see. Pluto is half the size of
Mercury and 1 billion miles further out than Neptune
(on average). The ETX-90 models just do not have the
aperture to even show this faint and small planet; its
magnitude is fainter than the limiting magnification of
the ETX-90 and its disk is smaller than the resolving
power. So, do not even bother trying to see Pluto unless
your telescope can see objects as faint as 14th
magnitude and resolve objects as small as 0.25 arcsec.

Asteroids

For many, asteroids are the forgotten objects in our
solar system. These are small objects, also known as
"minor planets", with diameters ranging from less than
a mile to several hundreds of miles, and are in orbits
around the Sun that are positioned mostly between the
orbits of Mars and Jupiter. Most of the many thousands
of asteroids are small and faint and therefore beyond a
small telescope. However, some of the larger ones can

be seen as star-like objects of magnitude 8–11. Many asteroids are in orbits where their apparent motion is quite large. This means that their changing position against the background stars is readily noticeable from one night to the next. Plotting the movement of such asteroids is a fun way to experience the dynamic nature of our solar system. Web Sites such as Heavens-Above (http://www.heavens-above.com) have current finder charts on asteroids.

Comets

Comets are occasional beautiful visitors to the inner solar system. You do not need a telescope to appreciate a comet and see the long "tail" that many comets have. But not all comets are the showcase objects that Comet Hale–Bopp was in Spring 1997. Most are moderately faint, and if any tail is visible it will be short. These comets, and all comets when far away from the Sun, are telescopic objects. You can use a small telescope to view these visitors, with the best views typically being had with minimal magnification. Most comets will have a "head" and a "tail". Inside the head will be the brighter "nucleus". As the comet approaches the Sun, the head and nucleus usually become brighter and the tail longer. As with asteroids, plotting a comet's movement against the background stars is fun. With some comets, movement can be noticed within just a couple of hours. The Harvard–Smithsonian Center for Astrophysics Web Site maintains an excellent listing of currently observable comets (http://cfa-www.harvard.edu/iau/Ephemerides/).

Amateur astronomers discover most comets. Some comet discoveries are made by dedicated comet hunters with specialized equipment and techniques, but other discoveries are made accidentally by amateurs with small telescopes who just happened to be looking for something else. For example, you may be "star hopping" (see Chapter 3) searching for a faint Messier object; but before you reach the final position for the object you want you see, you see a faint object that should not be there. You recheck your star charts and are convinced that there is no "faint fuzzy" object at that position. You then watch the object for a couple of hours and notice that it moves against the background of stars. You may have just discovered an unknown comet. Of course, before you report it you should check to confirm that the object is not really a known comet. Once you are certain you have discovered a new comet, you should report it to the International Astronomical Union Central Bureau for Astronomical Telegrams (see http://cfa-www.harvard.edu/iau/HowToReportDiscovery. html for details on the reporting process). Even though the comet you have just discovered currently appears as a faint fuzzy, it could become the next Comet Halley or Comet Hale–Bopp, except that it will be Comet "your name", and it and you will be famous! Good luck.

Meteors

Meteors are not normally observed through a telescope. On most nights when you stay outside for hours and take the time to look up at the sky (as opposed to looking into the eyepiece), you will see one or a few streaks of light. The light is caused by a small (usually dust-sized) particle entering the Earth's atmosphere and burning up. As the object increases in size, the brightness and length of the light streak can increase, sometimes dramatically. Occasionally a meteor will leave a trail of "smoke" in the atmosphere. This trail will change over time (due to upper level winds in the atmosphere) and eventually fade from view. Turning your ETX on to a meteor trail can present a fascinating view as the trail distorts, many times within minutes.

Stars

Many casual amateur astronomers do not consider observing stars as something worthwhile. After all, stars, extremely distant objects that they are, just appear as points of light regardless of the size of the telescope and the magnification used. While it is true that no details are visible when observing stars, this does not mean that they have nothing interesting to see. Many stars, both bright and faint, have observable colors. Even with small telescopes, the color of some stars is very evident. Blue (typically young stars), red (typically old), and yellow stars (middle-aged like our Sun) are star colors that are easily seen with the eye.

Famous stars such as Sirius (the brightest star in the night sky), Betelgeuse (yes, it really is pronounced as "beetle juice"!), and Proxima Centauri (the closest star to the Earth other than our Sun) are worth looking at just so that you can say you have seen them in your telescope. Other stars are "sign posts" on your way to observing other objects (using the "star-hopping" technique mentioned in Chapter 3).

While you navigate around the night sky with your telescope, do not forget to just stop and look at the stars in the eyepiece. Keep in mind that you may not be seeing just a star, but another solar system (even though you cannot see its planets) with all the possibilities that implies. Who knows, maybe there is someone there looking back at our Sun, wondering if there are planets orbiting that star.

Variable Stars

You may think that stars are unchanging, but you would be wrong. Many stars change their appearance. Sometimes this change is a result of their aging (they expand and contract and become brighter or fainter), and it can take decades and longer for changes to be apparent to the eye. However, some stars change their brightness on a regular basis that can be measured in hours or days. These are known as "variable stars" and are another type of object that demonstrate the dynamic nature of the Universe. Today, most observations and measurements of variable stars are done with electronic instruments, but the small telescope observer can still monitor the brightness changes of variable stars. You do this by comparing the star's visual magnitude with that of other stars in the eyepiece field-of-view. Doing this comparison over hours and days can reveal that the variable star's magnitude is actually changing.

To learn more about variable stars, check observing guides, astronomy software, or astronomy Web Sites (such as the American Association of Variable Star Observers, http://www.aavso.org/).

Sometimes a star increases its brightness in a dramatic fashion. In fact, the star "explodes" (expands in size and brightness). This event is known as a "nova" or occasionally "supernova". When a star will explode is unpredictable but when one does, the star can go from being invisible to visible or if previously visible, can change the look of the constellation it is in. Although rare, sometimes a supernova can even make the star visible in the daytime. Being the first to see a nova or supernova is possible for amateurs since there are so many of us observing the night sky. All that is required is that you know the night sky well enough to notice a star that appears "wrong". If you think you have discovered a nova, check the star charts to verify it is not an asteroid, planet, or known variable star. If you remain convinced that the nova is real, report it to the International Astronomical Union Central Bureau for Astronomical Telegrams (see http://cfa-www.harvard.edu/iau/HowToReportDiscovery.html for details on the reporting process).

Double Stars

Double stars are favorite objects for many small telescope observers. There are two types of double stars. With the first type, the two (or more) stars are just in proximity to each other in the eyepiece. With the other type, the stars (two or more) are actually physically associated with each other in some way, such as orbiting about the same common point.

Some double stars present observational challenges to the amateur astronomer, since their separation from each other is close to the resolution limit of the user's telescope. Ideal observing conditions and a steady mounting is required to separate these stars into the individual stars.

Other double stars are enjoyable to look at as each star may have a different color. One such striking example of this color difference is Albireo in the constellation of Cygnus (Figure 4.15).

Albireo is a fine double star; the star on the left in Figure 4.15 is a distinct yellow and the star on the right is blue. M40 (in Ursa Major) is a close (challenging to split) double star. See star catalogs for a listing of double stars.

Star Clusters

Star clusters come in all sizes and "shapes". Some have a few stars and many have thousands of stars. Some

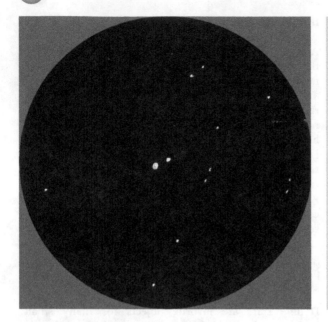

Figure 4.15. Albireo (Beta Cygni) in the 26 mm eyepiece.

clusters are called "open star clusters", because the stars are loosely scattered around in space. Others are called "globular star clusters", as their stars are tightly packed into a small "globe" of space.

When viewing star clusters you have to keep in mind the type, size, and overall brightness of the cluster in deciding how best to view it. Large open clusters are best observed with a low power and a wide field. Many open clusters will look beautiful in a small wide-field telescope like the ETX-70AT but be somewhat disappointing in a larger telescope like even the ETX-90, with its longer focal length and smaller field-of-view. The problem with the limited fields-of-view when observing such open clusters is that you can only see a few of the stars of the cluster at a time. Three such large open clusters are M45 (the Pleiades, seen in the photograph in Figure 4.16), the Double Cluster in Perseus (two large open clusters side-by-side), and M44 (the Praesepe or "beehive"). Viewing the Pleiades has been described as like looking at diamonds on velvet. Its bright blue stars really stand out against the dark background.

Other open clusters appear nice in a low-power eyepiece. A typical view you can see of smaller open clusters is shown in the drawing of M21 (Figure 4.17).

Other good open clusters are M6 and M7 in the constellation of Scorpius; M11 and M26 in Scutum; M18, M23, and M25 in Sagittarius; M29 and M39 in

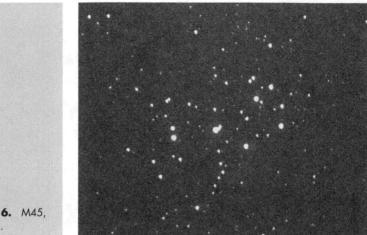

Figure 4.16. M45, the Pleiades.

Cygnus; M34 in Perseus; M35 in Gemini; M36, M37, and M38 in Auriga; M41 in Canis Major; M46 and M47 in Puppis; M48 in Hydra; M50 in Monoceros; M52 and M103 in Cassiopeia; and M67 in Cancer.

Globular clusters are much more compact than open clusters. The stars in these clusters are gravitationally bound tightly together. In most of these clusters, you will not be able to resolve the individual stars at the center of the cluster. With smaller globular clusters, you

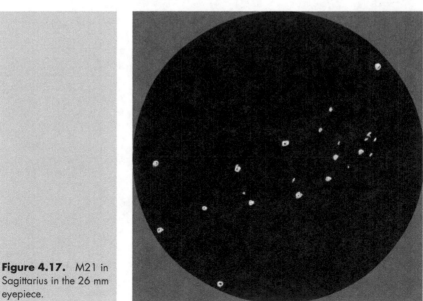

Figure 4.17. M21 in Sagittarius in the 26 mm eyepiece.

Figure 4.18. M22 in Sagittarius in the 26 mm eyepiece.

may not be able to resolve any individual stars at all, making the cluster appear faint and fuzzy just like a nebula (or gas cloud). However, there are many fine examples of globular clusters that are excellent objects in the ETX. Two examples are seen in the drawings in Figure 4.18 (M22) and Figure 4.19 (M13).

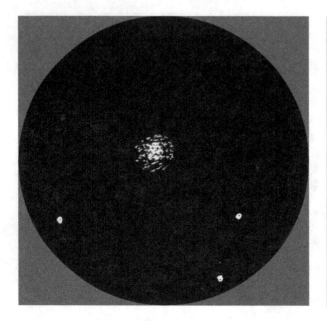

Figure 4.19. M13 in Hercules in the 26 mm eyepiece.

M13 is also known as the "Great Hercules Globular Cluster" and is a fine telescopic object, with many resolvable stars.

Other good globular clusters are M2 in Aquarius; M3 in Canes Venatici; M4 and M80 in Scorpius; M5 in Serpens; M15 in Pegasus; M28, M54, M69, and M70 in Sagittarius; M53 in Coma Berenices; M56 in Lyra; M68 in Hydra; and M92 in Hercules.

Some clusters of stars are not known as open clusters or globular clusters but are wide expanses containing hundreds or thousands of stars. One such object is M24 (the Small Sagittarius Starcloud),

Use the example drawings in this section only as indicators of the type of object you will be observing. Every cluster whether open, closed, or a star field will present different views. There will be different counts of stars, different patterns, different colors, different star magnitudes, and different star densities. Each cluster has a beauty of its own to be appreciated.

The Faint Fuzzy Blobs

Viewing nebulae (gaseous clouds) and galaxies is usually a disappointment to new amateur astronomers. The photographs in magazines and on the box the telescope came in look so nice, but the view through the telescope (any telescope) lacks the details, colors, and brightness seen in these photos. With some perseverance, nebulae and galaxies stop appearing as what I call "the faint fuzzy blobs" and become more examples of the wonders of the universe.

Nebulae ("nebulae" being the plural of "nebula") come in two flavors: diffuse and planetary. Diffuse nebulae are extended clouds of gas and dust with seemingly random shapes and sizes, whereas planetary nebulae have distinctive shapes, typically circular (spherical in three dimensions) and can look like small planets. Nebulae are also known as "reflection" or "emission" depending upon whether their light source is light from nearby stars or they emit their own light from the gas within the nebula (similar to the charged gas inside a fluorescent light).

First up, diffuse nebulae.

One of the first faint fuzzy blobs that most users will look at is actually not so faint, not so fuzzy, and certainly not a blob: the Great Nebula in Orion (M42).

Figure 4.20 M42, the Great Nebula in Orion in the 26 mm eyepiece.

This bright patch of nebulosity is visible to the naked eye in the middle of "Orion's sword". When you view it through a telescope more stars, dark lanes, and bright swirling clouds appear. The 26 mm eyepiece (or a low-power, wide-angle eyepiece) will show a lot of details in M42, if you spend the time studying it.

Near the center of M42, at the edge of some of the nebulosity, are four small bright stars in the shape of a trapezium (see Figure 4.20). This grouping is actually called the Trapezium. Depending upon the viewing conditions, only three of the stars may be seen; but the fourth is there. Larger telescopes (with more "light gathering power"), such as the ETX-125EC, show a larger expanse of the nebula and more dark lanes. In photographs, M42 is usually seen as reddish in color, but to the eye for most observers M42 will appear as pale green or perhaps even lack any color. M42 is both an emission nebula and a reflection nebula. Star birth is occurring in the Trapezium area even though you cannot see them "being born".

M8 (the "Lagoon Nebula") in Sagittarius (Figure 4.21) is another type of diffuse nebula presenting a less distinct shape to the eye.

Other diffuse nebulae that make fine but sometimes challenging objects for your ETX include M1 (the "Crab Nebula" in Taurus, actually a supernova remnant); M17 (the "Omega" or "Swan" nebula) and M20 (the "Trifid

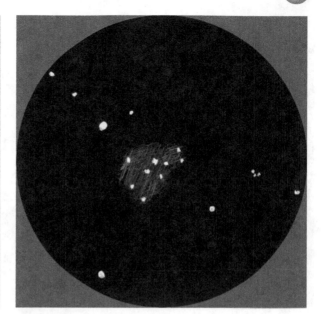

Figure 4.21. M8 in Sagittarius in the 26 mm eyepiece.

Nebula"), both in Sagittarius; and M43 and M78 (part of the M42 complex in Orion). When trying to see diffuse nebulae, some of the techniques discussed in Chapter 3 will be required. For example, to see M1 from moderately dark sky locations, dark-adapted eyes and averted vision will probably be required. When you do finally see it, you will likely not be able to discern any real shape to it in small telescopes; it will appear as a faint fuzzy blob.

Now on to planetary nebulae.

The most famous of the planetary nebulae is M57, the "Ring Nebula" in the constellation Lyra. This small round nebula is easily seen in small telescopes like the ETX-90. When observed from dark skies under good conditions with moderate magnification, M57 looks like a smoke ring (see Figure 4.22). In small telescopes it will take some time and experience to be able to see the dark central portion, but it is there.

Another good planetary nebula to view is M27 (the "Dumbbell Nebula" in Vulpecula). Do you see a dumbbell or just a round shape?

One of the more challenging planetary nebula objects is known as the "Ghost of Jupiter" (NGC 3242). This faint disk-like object almost does look like a very dim Jupiter. Other challenging objects are M97 (the "Owl Nebula" in Ursa Major) and NGC2392, both faint and small planetary nebulae.

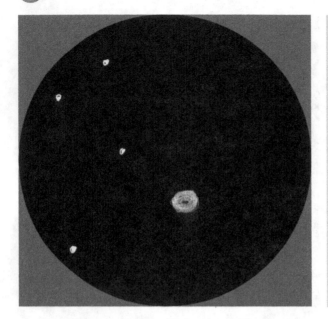

Figure 4.22. M57, the Ring Nebula in Lyra in the 9.7 mm eyepiece.

As with star clusters, each nebula will appear different. Some will be bright, some faint, some large, some small, some dense, some thin. To get the most out of observing nebulae, spend some time on each object you view, looking for as many details of it as can be seen. That is part of the enjoyment of "deep sky observing".

Galaxies

Up to now I have discussed objects that reside in our solar system or in or very near to our galaxy, the Milky Way. Now it is time to move outside our galaxy and view other galaxies, or "island universes", as they were once known. When you look at a galaxy, you are seeing hundreds of millions of stars in a very small space. Galaxies are so far away that light from most galaxies that you can observe with your ETX left its stars several million to tens of million years ago. Galaxies are classified into various types of "spiral" and "barred", but for most casual observers these distinctions are not observable.

Observing a galaxy requires the techniques mentioned in Chapter 3. Do not expect to see any spiral arms or "dark lanes" like you see in many photographs. Even though you cannot see details, it is thrilling to know you are seeing stars so far away, with perhaps worlds of their own.

The most famous galaxy (besides our own) is the Great Galaxy in Andromeda, M31. This bright and large galaxy (you can see it with the naked eye from dark locations) actually looks best in small wide-field telescopes (or even good binoculars). In the eyepiece of most small telescopes like the ETX-90, all you will see is an oblong fuzzy patch, which is actually just the central portion (known as the "nucleus) of the overall galaxy (see Figure 4.23). In fact, for most galaxies what you are seeing is the brighter nucleus area.

Many times you can make out the smaller and fainter nearby satellite galaxy of M31 known as M32. You may also see the second M31 satellite, M110. Other galaxies that are nice ETX objects include M33 (the "Pinwheel Galaxy") in Triangulum; M81 and M101 in Ursa Major; M83 in Hydra; and M102 in Draco.

More challenging galaxies include M49, M58, M59, M60, and M61 in Virgo; M51 (the "Whirlpool Galaxy") and M63 in Canes Venatici; M64 in Coma Berenices; M65 and M66 in Leo; M82 in Ursa Major; and M104 (the "Sombrero Galaxy") in Virgo.

Figure 4.23. M31, Great Galaxy in Andromeda, in the 26 mm eyepiece.

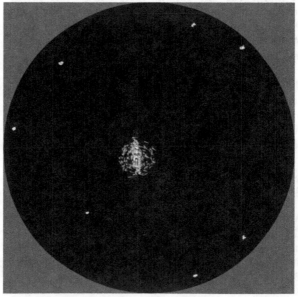

As with star clusters and nebulae, spend time viewing each object. Look for as many details (shape, size) as you can see. Enjoy the experience.

By the way, for much of this chapter, you have seen objects referred to as "M45" or "M31". Some new amateur astronomers may be wondering what these designations mean. The "M" refers to the "Messier" catalog of objects, logged by Charles Messier in the 18th Century. There are many catalogs of objects in the sky but the Messier Catalog is the most famous. There are over 100 objects in the Messier Catalog, all of which can be seen in a small telescope from a location with dark skies. You can learn more about Messier Objects in books or online at the Students for the Exploration and Development of Space (SEDS) Web Site (http://www.seds.org/messier/). Another standard catalog is known as NGC, or the New General Catalog. You will see many objects with their NGC number.

Man-made Satellites

Back a little closer to home, there are some challenging objects that the ETX can "see" in orbit above the Earth. Most of the brighter man-made satellites will appear as points of light in the eyepiece (a rapidly moving point of light). The one current exception to this is the International Space Station (there were two; Mir being the other one but it was de-orbited on 23 March 2001). When accurately tracked using a moderate power eyepiece, the ISS will show a definite shape.

Satellites are rapidly moving targets. You can track them by hand moving the ETX. Using the electric slewing arrows on your handcontroller will probably be an exercise in frustration since the satellite is moving in two directions (altitude and azimuth). Keeping the satellite in view will be a chase and a challenge. I suggest unlocking both axes and grabbing the tube. Keep the satellite in the eyepiece by moving the tube by hand.

If you have an ETX and Autostar, you can use the Autostar to locate and track satellites automatically. See Chapter 2 for information on entering satellite TLEs into an Autostar. This actually works extremely well and makes satellite chasing a fun thing to do with your ETX.

Terrestrial (Spotting Scope)

Getting very close to home, there are many things on the Earth that make great objects to view. The portability of the ETX, especially the ETX-60AT, ETX-70AT, and ETX-90 models, makes it an ideal "spotting scope" to take backpacking or on nature hikes. You can also use the ETX to view objects from your backyard or even from inside your living room. Normally, the manuals tell you to not use an astronomical telescope indoors, where air turbulence and window glass will deteriorate the image. That is true when viewing astronomical objects or when you want the clearest possible image. But if you just want to see something a little closer, using your ETX will bring it a lot closer.

So, just what can you look at with the ETX or other small telescope?

Airplanes are good objects to practice on while waiting for satellites to appear. Pick one several miles away or at high altitudes. Use different eyepieces to see more details or to provide more tracking challenges. It is a good test to see if you can make out the airline designation from the logo on the tail. Many decades ago when I was a teenage amateur astronomer, I would track nighttime military aerial refueling missions using a 3-inch telescope. It was fun seeing several airplanes so close together. Little did I know how close they were until I actually flew on aerial refueling missions as an Air Force fighter pilot!

If you live near the ocean or other large body of water, watching ships can be exciting. Perhaps you will be able to see some movie star on a cruise ship.

Even watching people (legally, that is) can be both fun and perhaps helpful. If you live near a park, you can watch your children playing (a great way to avoid the heat and noise). If you see a crime being committed, you could use your telescope to get a good look at the culprits or a vehicle license number and help the police make an arrest.

Buildings make nice objects; they do not normally move. You can try out your new telescope on a building, seeing the effects of image reversal and different eyepieces, and looking for any distortion in the regular lines of windows and other structures.

Using the ETX as a spotting scope to look at plants (the living kind, not the manufacturing kind) that are some distance away can be handy if you are a plant type of person. The high-quality optics will bring out colors and details just as you would see if you were standing close to the plant.

The high-quality optics in the ETX telescopes can also shine when viewing animals and birds. Many "birders" use the ETX-90 as a spotting scope.

You have spent a lot of money on a high-quality telescope. While you can get your money's worth from viewing the night sky, daytime use will add more value to your purchase. Do not just think of your telescope as a "night instrument".

Chapter 5

Photography

Many amateur astronomers quickly decide to try their luck with photography using their telescope. While this can be a rewarding experience, it is not without its challenges even when using telescopes designed for photography. With the ETX models, photography is certainly possible if you are willing to experiment (i.e., waste a lot of film) and are willing to devote a lot of time to its pursuit. Earlier in this book, you have seen some examples of astrophotography using the ETX. As long as you do not expect Hubble Telescope quality images nor 100-minute long exposures of a nebula, the ETX can make a good platform for certain types of astrophotography. This chapter will go into more detail on the photography that you can accomplish with a telescope that is not designed for astrophotography.

Before attempting any photography with the ETX (or any telescope, for that matter), there are some factors you must keep in mind. Small consumer telescopes typically do not come with very sturdy mounts. This can lead to vibrations and therefore blurring of the captured image. There are ways to minimize these vibrations (some are discussed later in this chapter), but the effects of a poor mount are difficult to overcome. Be certain you have a sturdy mount. If using a tripod, avoid extending its legs to any height beyond the lowest possible. This will reduce leg flexure as a source of vibrations. Using an alt/az mounting will also help, as this is a more stable configuration with the center of gravity of the telescope more nearly over the center of support of a tripod. Depending upon the weight of the camera, adding counterweights at the opposite end of the telescope may be required. This is done to keep the system balanced. It can also reduce the tension on the axis locks. Finally, the axis locking mechanism may be a serious factor. With some heavier cameras, the ETX locks are not sufficient to hold the telescope's orientation without slipping. This can make photography impossible. Proper counterbalancing can help but may not always work. Two such extreme examples are shown in Figure 5.1a and 5.1b. I attempted to use one ETX-90 piggybacked on another ETX. One ETX was used for visual guiding and a 35 mm camera was attached to the other ETX. This experiment did not work as the ETX locks were inadequate to hold the combined weights and the drive torque was insufficient to slew the system.

However, with some larger model telescopes, piggybacking a smaller "guidescope" is frequently done with good results.

Once you have compensated for its limitations, you can use your small telescope for astronomical photography or terrestrial photography. You can take really good photographs of the Moon, the Sun (with proper protections), some planets, and some other night sky objects. In terrestrial usage you can think of the ETX or any small telescope as a really long telephoto lens. When a 35 mm camera is mounted to the telescope without an eyepiece being inserted, the system acts as a telephoto lens having the focal length of the telescope. For example, the ETX-90 has a focal length of 1250 mm; so mounting a camera to the ETX yields a telephoto of 1250 mm, much longer than the typical 90–200 mm telephoto that many photographers may have. Along with its small size, this long focal length makes the ETX an ideal telescope for some type of terrestrial photography. I have received reports that many "birders" like it for its quality optics and portability.

The next sections discuss photography using different types of cameras. The accessories needed (if any) are discussed along with any techniques that can help improve your photographs. Most of the discussions pertain to astronomical use, but there is a lot that is applicable to terrestrial usage. The chapter will end with some simple techniques that can be used to improve the photographs that you have taken.

Before I move onto those discussions, a word about recording your photography sessions. Much of astrophotography is experimentation. Unless you know what

Figure 5.1a. Dual ETX system.

setups yielded what results, you can never hope to repeat the perfect shots you managed to get. You should record just about everything that can affect the photograph, from the equipment used, the configuration of the setup, the film speed and type, the object

Figure 5.1b. Dual ETX system.

photographed and anything special about it (i.e., the phase of the Moon), the exposure used, the techniques used to make the exposure (i.e., hat trick method, self-timer, etc.), and even the sky conditions. All of this information will be used after you see the final image. You can then document what configurations, film, etc. produced the best results. The next time you want to take similar photographs, you just refer to your notes. It may still be necessary to bracket or otherwise adjust some of the setup or exposure lengths, but at least you will have a good starting point.

35mm Film

For many years, amateur astronomers have used 35 mm cameras with their telescopes. The flexibility of these cameras makes them ideal for coupling to a telescope. Many have removable lenses, which is required for some types of mounting. The through-the-lens capability of a single lens reflex (SLR) camera helps when focusing or aligning the object in the camera. Many of today's highly automated 35 mm cameras are actually less desirable for astrophotography, which usually requires a great deal of manual exposure control. An older 35 mm SLR camera that has a "B" (bulb) shutter setting is more appropriate for use with the telescope. Two other features available in some cameras that come in handy with astrophotography are a manual flip-up mirror (to reduce vibrations induced by mirror movement) and an interchangeable view-screen. Most standard viewscreens are designed for well-illuminated scenes and do not perform well when trying to focus on faint astronomical objects. Some cameras can have the viewscreen replaced with a flat glass screen that works better in low-light situations.

With the ETX, there are four types of camera mounting that can be done. The first is "prime focus photography", the second is "eyepiece projection photography", the third is "afocal photography", and the last is "piggyback photography". I will discuss each of these.

Figure 5.2 shows a 35 mm camera mounted at the ETX prime focus position on the rear of the ETX-90RA. When the "flip mirror" on the ETX is rotated to this position (turned 90 degrees from that shown in the

Figure 5.2. Camera adapter and 35 mm camera at prime focus.

figure), light from the ETX passes through this opening and onto the film in the camera. The camera lens must be removed and no eyepiece is used in this configuration. The image is focused using the camera's viewscreen (SLR cameras only). This type of mounting requires two accessories: a T-Adapter designed for the ETX and T-Mount ring appropriate to your 35 mm camera. A locking flexible shutter cable is required for most cameras to reduce vibrations and to lock the shutter open (in the "B" setting).

In this configuration, you can take short duration (less than a second) photos of the Moon and the Sun (with proper protections). The Full Moon's image will nearly fill the entire height of a 35 mm film frame (as seen in Figure 5.3). When shooting the Moon and Sun, you may be able to get reasonably accurate exposure readings using a built-in light meter if it measures the light near the center of the field-of-view. You may have to decrease the exposure from that indicated if there is a lot of black sky affecting the reading. Using ISO 400 color film, I have taken good photographs of the Moon with exposures ranging from 1/30 second (partial phases) to 1/125 second (Full Moon). You will have to experiment with shutter settings to get the best shots. If you get a light meter reading, bracket the exposure one or two shutter speed settings on either side of this value. If you cannot get an exposure reading, then use more settings or start with these values (adjusted for your film speed).

While you can take photos of the brighter planets, their size on the film will be too small to be really

Figure 5.3. Full Moon using prime focus.

useful. You might think that you could take long-duration (several minutes) photos of nebulae and galaxies in this configuration. Although you can do this, the accuracy of the ETX polar alignment and tracking will probably prove to be less than perfect, resulting in serious trailing of the image on the film. To take long duration astrophotos, you need some way to correct the guiding as the exposure is underway. In this configuration, there is no way to do that unless you have what is called an "off-axis guider". This adapter allows a small portion of the field-of-view to be directed to the side where a guiding eyepiece is inserted. Most small telescope users do not attempt this type of long-duration astrophotography, although it can be done. With the ETX, there is a better way to get long-duration photographs of the night sky and that is by using the piggyback technique discussed shortly.

If you want to increase the size of the image on the film, you will have to use an eyepiece. This technique is called "eyepiece projection", since the image from the eyepiece is projected onto the camera's film plane. Figure 5.4 shows a 35 mm SLR camera mounted to the ETX using the Basic Camera Adapter from Meade Instruments. An eyepiece is inserted into the adapter and the adapter is inserted into the eyepiece holder on the ETX. The camera is attached to the adapter using a T-Mount ring.

Figure 5.4. Eyepiece projection camera adapter and 35 mm camera.

Figure 5.5. Craters along the Moon's terminator.

In this configuration, you can take photographs with more magnification, depending upon the focal length of the eyepiece used. However, there are some factors that limit this type of astrophotography. The most serious factors are mount stability and the reduced amount of light reaching the film. This type of photography will be most effective with photographs of the Moon, especially craters near the terminator, using a moderately high-power eyepiece (say, a 9.7 mm). An example of this is shown in Figure 5.5.

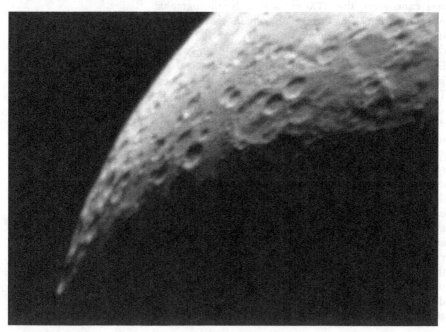

This is a 2-second photo on ISO 800 color film using the 9.7 mm eyepiece. With the longer exposures that are required at the magnifications available with eyepiece projection, any vibrations in the telescope will ruin the image. A sturdy mount and the vibration eliminating techniques discussed later in this chapter are definitely required for best results. As with prime focus photography, long-duration exposures (several seconds to several minutes) are not really possible in this configuration.

The third type of camera mounting is known as "afocal photography". In this mode, you leave the camera lens attached, meaning you can use this technique with almost any type of camera whether film or digital, still or video. You shoot through the eyepiece. In fact, you focus the eyepiece for your eye and then point the camera lens into the eyepiece and take the exposure. You can even handhold the camera over the lens for short-duration photographs of brighter objects. As this technique works really well with some digital cameras, it is discussed in more detail in the discussion of digital camera astrophotography later in this chapter. Since you are using an eyepiece, the same limitations apply as with eyepiece projection photography.

OK, so you cannot take long duration photos with the previously discussed mounting methods. Does this mean you cannot use the ETX to take such photographs? Absolutely not. Using any 35 mm camera that can have its shutter locked open, you can take excellent long-duration photographs of many sights in the night sky. Figure 5.6 shows a 35 mm camera with an attached 90–230 mm zoom telephoto lens attached "piggyback" style. The ETX is used to track the object's movement in the sky, thereby reducing or eliminating trailing on the film.

Except for very short-duration exposures, it is necessary to correct for tracking errors by manually guiding the telescope. For ease of tracking on a "guide star", use an illuminated reticle eyepiece. This type of eyepiece has a grid of lines etched onto a surface that is illuminated by a small light source. Using the grid, it is easy to keep a guide star in the same position within the eyepiece field-of-view. If you do not have such an eyepiece, use the highest power eyepiece with which you can comfortably observe and track a guide star in the center of the portion of the sky you are photographing. Use the right ascension and declination drive motors (at the slowest speed) to keep the guide star at the same position within the eyepiece. Do not

Figure 5.6. Camera and telephoto lens piggybacked on ETX-90RA.

Figure 5.7. Piggyback photography of the heart of the Milky Way.

worry about small errors; as long as you keep the star position reasonably constant during the exposure, the effects of any movement on the film will be negligible. Using this technique you can take long-duration photographs showing constellations. When using color films, you can even discern star colors. With a normal 50 or 55 mm lens on the camera, you can obtain beautiful photographs of the night sky. An example is

shown in Figure 5.7. This 15-minute exposure of the heart of the Milky Way was taken with ISO 800 color film using a 55 mm lens. I used an illuminated reticle eyepiece to manually correct for tracking errors. The dotted streak along the lower right-hand portion of the photo is a passing airplane.

If you use a telephoto lens on the camera, you can get higher magnification astrophotographs, like that shown in Figure 5.8. This 10-minute exposure of M42, the Great Nebula in Orion, was taken on ISO 800 color film using a 230 mm telephoto lens.

Except for the ease of digital camera photography (discussed next), long-duration piggyback astrophotography is likely to be the most rewarding type of photography to ETX users.

In any 35 mm photography, when the exposure time is more than a few fractions of a second, vibration from any source will result in blurred images on the film. Wind is one source of movement of the telescope. Attempting long-duration astrophotography in a strong breeze requires a sturdy mount and perhaps even some protection from the wind. For best results, do not try it when there is any breeze blowing at the telescope's location. In many 35 mm cameras, the movement of the mirror (SLR cameras) and the shutter can cause significant vibration to be transferred to the telescope. Use of the "hat trick" method is a way to eliminate this source of vibration.

Figure 5.8.
Piggyback photograph of M42, the Great Nebula in Orion.

Hat Trick Method

1. Mount the camera as required on the telescope. Point the telescope at your target object. Verify the position of any guide star in the eyepiece.

2. Next cover the lens of the camera or telescope with a dark object. Do not touch the lens or telescope; just block any light from entering the optics.

3. Set the camera to its "B" (bulb) setting and lock it open. Allow any vibrations from mirror/shutter movement to dampen out.

4. Being careful not to touch the telescope or lens, quickly move the cover out of the light path. Let the exposure go for as long as necessary.

5. At the end of the exposure, move the cover back into the light path and release the camera's shutter lock to close the shutter.

While this technique is ideal for long-duration photos (many seconds at a minimum), it can also be used for moderately short-duration exposures. With some practice, you can get exposure times down to less than a second. The only problems with this technique for short exposures are accuracy of the exposure length and repeatability.

When using film for your astrophotography, unless you have a photographic darkroom at your disposal, you may find that the commercially made prints are not exactly what you expected. If you have a flatbed or 35 mm film scanner, you can scan the images into your computer and manipulate them as described later in this chapter. Alternately, you can have your photos put onto PhotoCD or Picture-on-disk at the time they are developed. PhotoCD provides your photos in higher resolution (actually several resolutions up to 3072 × 2048) than Picture-on-disk where they are normally lower resolution JPEG files. Currently PhotoCD supports up to 100 images on a single CD-ROM disk. The PhotoCD laboratory can add each roll of film as you take them until the 100 image limit is reached. There is no need to wait until you have 100 images available for processing.

Digital Camera

When I first purchased an ETX in September 1996, I began to experiment with digital camera astrophotogra-

phy. Using my first digital camera, a Casio QV-10 (one of
the first consumer digital cameras), I was amazed at the
images that I could capture through my ETX. Using the
afocal photography method, I took photos of the Moon,
including a lunar eclipse, and the brighter planets. I later
used this camera for solar astrophotography (using a
solar filter) and obtained some nice photos of sunspots. I
even took some photos of an occultation of Saturn by the
Moon (see Figure 5.9).

Figure 5.9.
Occultation of Saturn.

With today's digital cameras, it is possible to acquire
amazing, quality images of many astronomical objects.
Shooting the Moon is easy, but some users are
obtaining nice shots of star clusters. All it takes is
some patience and knowing your camera's capabilities
and limitations.

Until you learn how to use your camera/telescope
combination, I suggest practicing (a lot) on the
Moon. The simplest way to use a digital camera with
your telescope is to focus a low-power eyepiece to
your eye. Then hold the camera over the eyepiece
with its lens directed at the eyepiece. Look at the
image on the camera's LCD display. If all you see is a
small circle of light from the Moon's image, you may
need either to zoom the camera lens to increase the
size of the image on the display or use the macro
mode (if the camera has this mode). Once you have
the image reasonably sized on the display, hold the
camera very steady and trip the shutter. If the auto-
exposure worked acceptably, you will get a properly
exposed photograph of the Moon. If the Moon is
overexposed (a likely possibility), you may have to
adjust the camera's controls to reduce the exposure.
Using a Moon Filter on the eyepiece can also help.
After you have some experience at getting good
photos at low magnification (you will probably be
deleting a lot of photos from the camera's memory!),
you can begin to experiment with higher-power
eyepieces. As the magnification is increased, the
surface brightness of the object is spread out across a
larger area, thereby increasing the length of the
exposure required. At some point, the exposure will
be too long to handhold the camera without blurring
the image. It is at this point that you realize you need
to attach the camera to the telescope using some sort
of adapter.

Figure 5.10a shows a digital camera attached using
the Scopetronix (www.scopetronix.com) Digital Camera Adapter. The digital camera is shooting through

an eyepiece mounted at the ETX rear port using the Shutan (http://www.shutan.com) Wide Field Adapter (focal reducer). Figure 5.10b shows a digital camera "hard-mounted" to the eyepiece using a Scopetronix Digi-T System. The Digi-T combines an eyepiece and camera into a single device and puts the camera lens very close to the eyepiece, reducing vignetting and eliminating stray light entering the camera. When a digital camera is mounted above or on the eyepiece, there is no longer a need to handhold the camera and so there is no camera movement. However, it is still possible to induce some vibrations into the system when touching the shutter button. To avoid this source of vibration, use the camera's self-timer mode. If the length of the delay is adjustable, set it for at least 10 seconds. Align and focus the image on the display, enable the self-timer, and then press the shutter release.

With experience you can capture amazing images of the Moon, the Sun (with proper protections), the brighter planets (showing the phases of Venus, cloud bands on Jupiter, and the rings of Saturn), and perhaps even more. And as the film is "digital", you can delete the bad images before anyone but you sees them.

Video

Some ETX users have used a video camcorder to capture what the eyepiece "sees". For bright objects, the camcorder can be handheld. For best results, the camera should either be attached with an adapter or placed on a tripod positioned next to the ETX. If attached directly to the ETX, the weight of the camera may become a problem if it is too heavy. To use a video camcorder, you use the same techniques as discussed previously for a digital camera in afocal photography mode. One user submitted an impressive example of a lunar video to my ETX Web Site. He had captured the passage of an airliner in front of the Moon while taping the Moon through an eyepiece. Do not dismiss this type of photography as being impossible or too difficult. Today's video camcorders can perform well for some types of astrophotography.

At the other end of the video camera spectrum are the desktop video cameras such as the "QuickCam".

Figure 5.10a.
Digital camera adapter
holding a camera at the
eyepiece.

Figure 5.10b. Digi-T
camera adapter holding
a camera at the
eyepiece.

Many users have used this simple video camera to capture images through their telescope. In some cases they have used special software to combine multiple images into a single image to produce excellent photographs of the Moon and even some planets. While this camera is not sensitive enough to capture faint objects, it can be successfully used on brighter objects. For best results, the camera should be attached to the eyepiece using an adapter. For more on using a QuickCam or similar desktop video cameras, see my ETX Web Site.

CCD

A step up from digital and video cameras is a CCD (charged coupled device) imager. These devices are more sensitive and normally more expensive than digital cameras or the desktop video cameras. In addition they come with specialized software to capture and perhaps integrate (combine) images to produce a high-quality image. One such CCD that has become popular with the ETX community is the Sonfest Promotions (http://www.sac-imaging.com) SAC CCD imager. In the hands of experienced users, this low-cost CCD system has been yielding very high-quality images of the Moon, the brighter planets, and even some fainter objects. Many examples are available on my ETX Web Site.

Improving the Photos

Using software you can enhance the appearance of your photographs. This can be done with Adobe Photoshop, software that comes with some scanners and digital cameras, or even with some shareware applications. While I have Adobe Photoshop, I actually use the Macintosh shareware application GraphicConverter ($35) from Lemke Software (http://www.lemkesoft.com) for most of my graphics manipulations and conversions. Figures 5.11 through 5.16 show some before and after examples of photographs of the Moon taken with a Ricoh RDC-4200 digital camera. The photos were taken at a

Figure 5.11. Before enhancement.

Figure 5.12. After adjusting "levels" and sharpening.

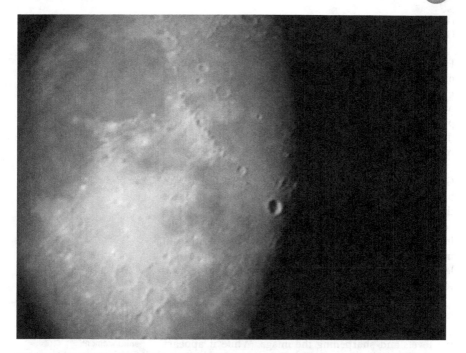

Figure 5.13. Before enhancement.

Figure 5.14. After enhancement.

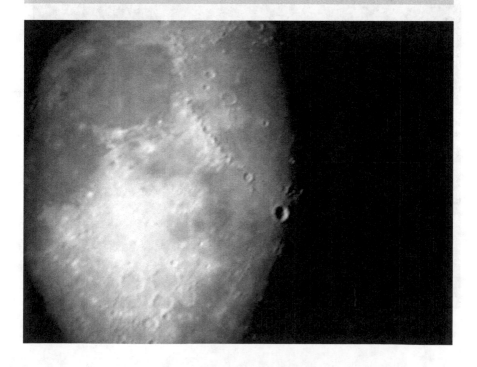

resolution of 1280 × 960 pixels and reduced to 33% in GraphicConverter.

Figure 5.11 (southern region of the Moon just past First Quarter) was taken using the 26 mm eyepiece. In order to get a large image on the camera's LCD display, I zoomed the camera lens to its full 3× position. This was necessary to get the camera lens close to the focal position of the eyepiece. The camera was handheld over the eyepiece (afocal photography) and the shutter released.

Figure 5.12 shows a drastic improvement. Using GraphicConverter I first adjusted the "levels" to brighten the Moon's surface. This compensated for the slight underexposure. Then I used the "sharpen edges" command to increase the contrast across the image. This has the effect of making edges stand out more than in the original image. Use caution since both levels and sharpening can be overdone. Neither can totally compensate for a bad photograph.

Figure 5.13 (central region of the Moon) was also taken using the Ricoh and the 26 mm eyepiece.

Figure 5.14 shows the enhanced image after adjusting the levels and sharpening the image. While it appears

Figure 5.15. Before enhancement.

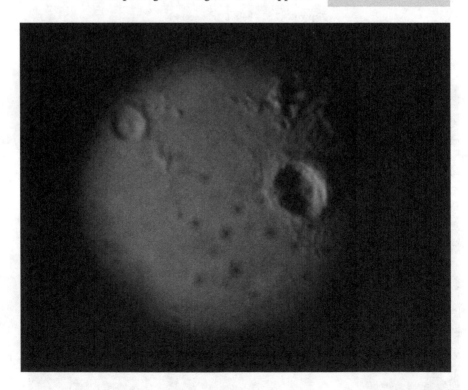

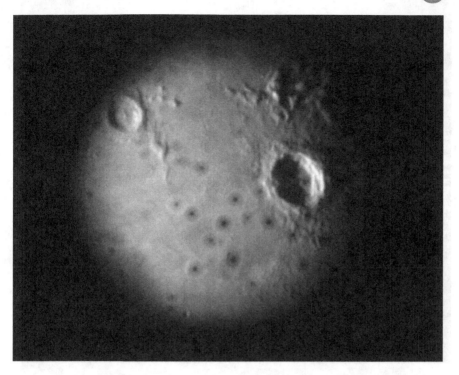

Figure 5.16. After enhancement.

that new details are visible in the enhanced image, this is not the case. Only the contrast between pixels makes it appear that way. The large crater near the terminator is Copernicus.

Figure 5.15 (the crater Copernicus) was taken through a 9.7 mm eyepiece. In order to get this photo, the camera was mounted over the eyepiece using the Scopetronix Digital Camera Adapter and the camera's self-timer used to allow vibrations to dampen out after pressing the shutter release.

Adjusting levels and sharpening resulted in the seemingly amazingly clear photo of Copernicus with some details in the crater wall visible (Figure 5-16). These images show the severe vignetting (image not filling the frame) that can result with some eyepiece/camera combinations. The images also show the results of a slightly dusty eyepiece! Using software (and a lot of work) it is possible to remove these circular spots.

You can go another step and even remove the artifacts, ending up with something like Figure 5.17, which is an even better rendition.

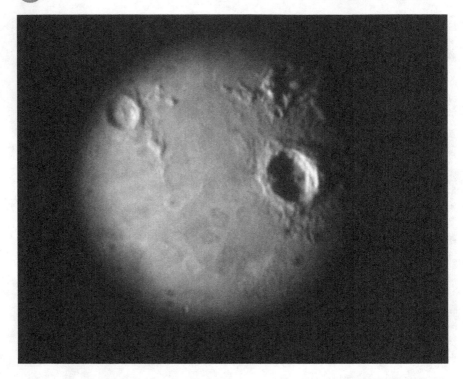

Experiment with your images and software until you get the results you like. Try using other adjustments (saturation, contrast, etc.) as well. Always work on a copy of the file so that you can start over after totally messing up the image.

Figure 5.17. Spot removal.

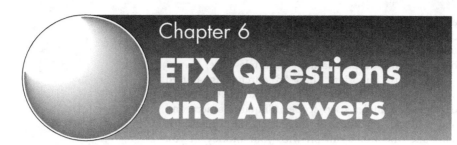

Chapter 6
ETX Questions and Answers

Since its creation in September 1996, the Mighty ETX Site on the Web has become a respected resource known worldwide for its tips, information, how-to's, and actual ETX astrophotography and as a forum for questions and answers from real ETX users. For readers of this book who are not "Net-connected", this chapter will give you a flavor of the ETX Site as well as provide some valuable information about using and improving your ETX. For readers who do have access to the Web, this chapter will provide in more permanent hardcopy some information from the Site that has proven very popular. I wish that the entire Mighty ETX Site content could be printed here; but as of July 2001, according to one visitor who printed everything, over 3000 pages came out of his printer!

Purchasing a Used Telescope

When Meade released the ETX-90EC, there were a lot of the original model ETX telescopes for sale by those wanting to upgrade to the Autostar-capable model. Then when Meade released the ETX-125EC, again there were older models for sale by those wanting the larger aperture and longer focal length. However, for many users these older telescopes, now available at a reduced price, were just what they wanted to get started in amateur astronomy. But these purchasers were rightly concerned about buying a used telescope. In response to a frequent question on the ETX Site, I posted this basic guidance on purchasing a used telescope:

"The most important thing in purchasing a used scope is to buy it from someone you trust, either a dealer who handles used scopes (Shutan Camera and Video does, or did, and there may be others) or someone who won't run off with your money. If you can't see the telescope in person (such as when purchasing via eBay), you may be taking some chances, unless you purchase from a reputable dealer. If you can see the scope in person, check the optics for dirt, scratches, etc. Take it outside at night and do a star test (see if the circles of light are circular and concentric as you move from one side of the focus to the other). Check the physical condition of the tube, fork mount, finderscope, and accessories. Dents and dings can be a sign of rough treatment and could result in problems. Look inside the battery compartment for signs of corrosion. Check the tripod mounting holes (all of them) for stripped threads. Lock both the RA and DEC and check for excessive play. Unlock them and check for smooth movement. Slew the scope electronically and listen to the drives; there should not be too much variation in the sound level or pitch. With either the standard handcontroller or the Autostar, check that all the buttons are functional."

If you follow this guidance, you will improve your chances of being satisfied or at least not too frustrated with your purchase of a used telescope.

Whether purchasing a used or new telescope, you will find a lot of good information on the "Buyer/New User Tips" (http://www.weasner.com/etx/comments.html) page on the ETX Site.

Telescope Maintenance

One of the many valuable tips on the Site for new telescope owners (and for long-time owners who've become lazy) is how to maintain the beauty and functionality of their ETX. The following article (courtesy P. Clay Sherrod) details some important considerations and techniques for maintaining and caring for the ETX.

Introduction

Is your ETX telescope "Sexy?" Does it look like mine? When I first got my ETX-125, I vowed to never hang every gadget on one end or the other. I wanted my telescope looking just like it did when I got it: shiny blue with lines as sleek and smooth as a formula racing car.

You have to admit they are a pretty sight when they are brand new. Face it, part of the reason that you bought the ETX is because ... well, because it's "sexy."

Well in my case, eventually nature called (literally in this sense); and now my telescope weighs probably twice what it did when I got it, with nearly every spare inch of available space holding dearly to some doodad that I simply could not go any longer without. I have finders on both sides of the eyepiece, clamps everywhere and cords going to God only knows where.

Now, on my once clean heavy-duty tripod are power cords, inverters, converters, reverters, and diverters (just kidding about the last two ... please don't ask what they are, because I know you'll want one). I have plugs for AC and DC and power switches for every direct voltage from 1.5 to 18 volts. There are little light bulbs under the wedge so I can see where the telescope is going, on the pier and tripod so I can see where my accessories are going, and even on the ground so I can see where I'm going! (See Figure 6.1 for Clay's telescope.)

Amateur astronomers are a funny bunch. We think we have to have one of everything that shows up in some advertisement. We are sort-of a high-tech, brainy keep-up-with-the-Joneses bunch.

I remember when I first logged onto Mike Weasner's ETX Web Site only to find a report from a nice fellow who had changed over to the Heavy Duty tripod for his ETX-125. His wife told him it really looked "sexy," which made the guy really proud of the scope and helped her by keeping him away from yet another larger telescope. Perhaps that's what it's all about – performance ... and sex appeal.

Well, I'm proud to say that even with all the junk hanging on my scope (and trust me, I use every bit of it!) my scope still looks "sexy," and I'm going to keep it that way.

Keeping your telescope in top shape is very important. I plan to keep mine (it has retired with me) and will do everything I can to keep it in as good a shape as I possibly can. Let me share with you a few

Figure 6.1. Clay Sherrod's "sexy" ETX-125 awaiting any observing requirement that it may face. The tube is cleaned after each observing session with a damp cloth and the mounting/base are protected with Armor-All automotive protectant against moisture, fading and fingerprints. Courtesy P. Clay Sherrod.

quick tips on keeping your telescope "sexy" for many, many years to come.

Use of the Telescope Outdoors

Telescopes are made to be used outdoors obviously. But even made that way, the worst thing that can happen to them (unless you are careless) is the effect of outdoor conditions.

Moisture is the biggest natural enemy of a telescope. Temperature is the second. Human use is a close third to numbers One and Two. Let's talk about moisture in two ways: outdoor and indoor.

Outdoor Moisture

Every time we take our scopes outdoors, we are subjecting it to moisture; I corresponded with one nice fellow from Scotland who actually observed the Moon (for a project) while it was raining on his balcony. Not a particularly good idea for most of us. Don't do that.

Normal moisture collecting somewhere on the telescope will happen nearly every time you take it out. You will have either dew or frost; they are both the same and form when the air temperature drops to, or below, the "dew point" that the weather man always mentions on your local forecast. Expect dew or frost; it's one of those necessary evils of the hobby. Note that dew nor frost rarely form when a brisk breeze is blowing at night.

Below are some common-sense care tips for dealing with outdoor condensation:

1. Never wipe off your optics, no matter how much dew or frost gets on them; bring your scope in or cover it up (with a pillow case or sheet) if it gets that bad.

2. Always monitor your objective lens to see if dew is forming; never let it get so bad as in (1) above. The best way to monitor this is with a flashlight aimed across, not directly in front of, the glass; if you see a cloudy film, then you may as well quit unless it is a special event such as an eclipse.

3. If the outside parts (the fork arms, tube, tripod, etc. – nonoptical) get moist, don't worry about it until you bring it in. You'll drive yourself crazy wiping it off.

4. Keep eyepieces covered in their little cases until ready to use, and once done replace them back into the cases.

5. All of your charts and sky maps should be covered for protection from dew as well; they will form dew more quickly than your telescope.

6. Keep moisture away from all electronic components. Your handcontroller, a DC or AC inverter if you are using such, and even the electrical plug connection should be raised above ground level if the grass is beginning to get wet.

7. "Parking the telescope": If you are at a star party, camping, or even at home and know the weather is going to be nice again tomorrow night and do not want to bring the telescope indoors, follow these rules to protect the telescope:
    (a) If it is not going to be raining, or if the winds are not excessive, it is perfectly okay to leave the scope out, provided you have run all the burglars away first.
    (b) Make sure your power connections are undone and your off–on switch is "off".

(c) Cover the telescope with a soft, clean pillow case or cotton sheet.

(d) Cover that with a small plastic tarp with elastic "tie downs" to something firm on the ground; do not tie the plastic tightly around the telescope or moisture will condense during daylight hours!

(e) Uncover the telescope about one to two hours prior to use to equalize for the evening.

Indoor Moisture

Unless you live like Snuffy Smith and have holes in your roof, you should not have to worry about indoor moisture. However, as discussed in a moment, lack of moisture in your home or storage area for your scope can be very detrimental to the telescope, particularly in winter months.

1. When bringing in the telescope after dew or frost has formed on it while observing, never cover up the front lens. Always plug up the eyepiece holder so that moisture cannot condense inside the telescope.

2. In winter months, only if moisture has not formed on your lens, cap it up securely (do not over-tighten the lens cap) and bring the telescope indoors with all optics covered. Condensation will immediately form on the outside of the telescope and mount; don't worry about this right now.

3. During other months, if moisture does form on the lens, bring it in uncovered and let it evaporate naturally and slowly indoors. Only then, after drying, if you can see any significant – and I mean significant – spotting from the moisture, do you clean the lens carefully using the cleaning method described on the ETX Web Site.

4. No matter whether the moisture forms outdoors or indoors, after bringing in the telescope, use a soft cloth (I prefer Terrycloth) towel and gently wipe down all metal and plastic parts until free of water.

5. Electronic components outdoors: Your electronic components are temperature-sensitive and must be protected from extreme heat and cold. In very cold weather, electronic handcontrols can do strange things and they should be kept warm whenever possible. Many people keep them (yes, they do) in small can coolers when not in use and carry them in their pockets if going inside for an extended time.

6. Optics on a hot day: If you are going camping or getting ready for a night-long star party and want to set up early, there is a very important rule: never leave your telescope in direct sunlight for a long period of time. Just as in a closed car, the inside of your optical tube assembly is capped off from ventilation and will become very hot. The baffle on the secondary uses adhesives to hold it in place, and there are cements used in various places throughout your telescope. Always protect it from heat outdoors and in an automobile. Use the child-care rule: "Would I leave an infant in conditions like this?"

The Telescope Indoors

To me, the worst treatment that a telescope gets is no treatment at all – not ever using it. This allows dust to accumulate. Yes, dust does settle inside the fork arms and the drive base, causing problems in motion over long periods of time.

In addition, a stored telescope tends to redistribute its lubricants (on the drive gear, the bearings, the fork arm drives in the ETX-EC, and even the focus mechanism) when it sits in one position for a while; gravity, a very strong force over time, will take the lubricant and put it where it wants it. You are left with a very dry drive system, except in one spot.

Even if you do not use your telescope to observe with for an extended period of time, go in where you have it stored (see below for telescope storage) and move it around occasionally. Turn on the motors and slew around. Focus on nothing in particular. Let the telescope know it is still loved.

Dust

Dust needs to be kept off your telescope; you bought a beautiful instrument: keep it that way. Remove dust gently and always with a soft damp cloth; do not use *Pledge* or any other dusting compound. Use water, and only a little bit of that. Never use window cleaner on any part of your telescope! Use only the optical solution described on the ETX Web Site for your optics and only use water for the rest of the telescope. Water will restore the brilliant shine and color to your tube assembly.

Once the dust has been removed from the telescope, gently buff it with a soft towel to make it shine like new (I use an old diaper).

Tripod

If your tripod is the painted type, as are most camera types and those that are commonly used with the smaller ETX scopes, simply clean them with water as well. However, if you have the heavy-duty chrome tripod with steel legs, you need to use automotive chrome cleaner to periodically provide a protective finish to the legs. All exposed metal parts, whether cast aluminum or steel (like the hardware), should be sprayed with a light coat of WD-40 and wiped down to an even shine for protection. Do this a minimum of once a month.

Fork Mounting

My fork mount still looks "sexy" after all these nights out. How do I keep it looking that way? Armor-All. That's right, Armor-All, the "car stuff." In case you haven't learned by now, the ETX is prone to fingerprints. What you may not know so far is that the plastic, just like the dashboard of your car, is also prone (very much) to solar fading over time. This is particularly true, not just from exposure to the Sun but also from storing the telescope under fluorescent lights. If you wonder why the fork assembly has lost its luster and now appears a rather ash gray instead of black, that's probably the culprit.

I wipe my ETX fork arm and drive base (including the setting circles and clamps) thoroughly with two coats of Armor-All every week. It pays off, if you want a sexy scope that you are not ashamed of in public. The first coat, I guarantee you, will soak right in. Let it dry hard, and then buff to the best shine you can. Always buff with a clean, dry, soft cloth. Then repeat by applying another coat of Armor-All (or other similar product) onto the plastic components; let dry thoroughly and buff again. You will be amazed at how good the telescope turns out – better than it did when you got it.

Not only does it look good, but you have protected the plastic and made it resistant to fingerprints as well.

Do not spray the solution onto the telescope; spray it on the rag or otherwise it may get down into the crevices around the setting circles and clamps; I always carefully wipe down everything black (except the lens cap, which is metal), including the rear cell of the telescope, the electric focuser, the right-angle holders for both finder and scope, the optical tube assembly support arms, etc.

Inside the Telescope

For the novice I do not recommend getting "inside" the fork arms or drive base of the telescope; if you must, for whatever reason, refer to the ETX "Tune Up" articles on the ETX Web Site. They thoroughly describe the process of lubrication and adjustment.

For most people, getting inside the telescope should never be necessary and maintenance is certainly not normally required. Never attempt to get inside the optical tube assembly except to clean the corrector lens and then do not be tempted to explore further.

I do recommend, however, even for the novice, periodically (say every six months) removing each of the side clamps on the altitude (declination) arms and thoroughly cleaning to remove the dust from all surfaces behind them. You must be careful when unclamping them completely, however, so that the optical tube does not fall down into the base. Remember that a small amount of lithium grease is required behind each setting circle where they rub against the four protruding plastic posts from the fork arm; otherwise you will restrict free movement of the assembly during electronic slewing and/or tracking.

Telescope Storage

There undoubtedly will be times when you must part company with the telescope for a while or periods in which you cannot use the telescope. Take the following precautions for its care during dormant periods:

1. Check to assure that all optics are covered; fumes from heating and air conditioning can leave very troublesome deposits on optics.

2. Cover the scope with a cotton pillowcase or sheet.

3. Do not cover that with plastic, as you did when "parking" the scope for a night.

4. Position the telescope so that no vents blow directly on it.

5. Never store the telescope in an open window no matter what direction it faces. If it faces east, south, or west, the sun can severely fade the finish (even covered) and the heat from it can damage internal parts. If you put it in a north-facing window, it is likely to develop mildew after an extended period of time.

6. Keep in mind that lack of moisture, just like too much, can damage your telescope. In this case, it can result in adhesives (such as that which hold the secondary baffle) drying out and cracking. Dryness can result in premature aging of the plastic, much as happens to automobile tires when they "dry rot"; and very dry conditions will definitely adversely affect any rubber or soft plastic fittings on your telescope or accessories.

That's it. Keeping your telescope clean is like taking a bath; you sometimes just hate to do it, but you gotta'. Otherwise, you might not be "sexy."

High-Temperature

Along with the previous article on maintaining your ETX, the dangers of excessive heat is expanded upon in the following article (courtesy P. Clay Sherrod).

Summertime ...

And the Livin' Ain't So Easy for your ETX Telescope!

We in the United States have been through one of the hardest winters on record and perhaps the first cold weather for most ETX users. Through this we have learned a lot about the computerized GOTO telescopes, a lot we wish we did not have to know:

1. Cold weather drastically affects the mechanical performance of the Meade telescopes because of the heavy green grease and lubricants that gum up in subfreezing temperatures. The only solution to that

problem is to "de-grease" (see the Performance Enhancement Guide Part 1, below).

2. The required 12 V power during cold months is adversely affected in very cold weather, usually beginning below 27 F and getting worse the colder it gets. After battery power gets too low, motor function and GOTO accuracy can become next to impossible.

3. The Autostar is most definitely temperature sensitive at the same cold range. It must be kept warm, and it must be supplied with no less than a constant 10 V. Otherwise the display begins to talk to you in "Martian" and the keypad commands to the ETX scopes become inoperative.

But we're through all that, right? Summer is coming, and the "... livin' is easy" – until the sun bakes your Autostar or until temperatures get so warm on your precious telescope awaiting nightfall on some August afternoon that the very glues that hold your secondary baffle in place give way, or your circuit boards no longer can "resist" nor "capacitate".

Welcome to the perils of summer and hot temperatures. We are in for another bout of temperature-related ETX frustrations and failures. I want to alert all ETX users to the dangers that can cause irreparable harm to their telescopes, the circuitry, and particularly to the Autostar – some of this has been based on actual tests that I have completed on standard #497 Autostars.

You are likely to be surprised, even if you are a seasoned "veteran" user of the ETX telescope and Autostar.

Your Telescope in Summer Heat – Part 1: Optical Tube Assembly, Mount, Accessories

Following is a quick checklist of (P) potential heat-related problems with the telescope and mechanical equipment itself and (S) the solution to prevent the situation from occurring, and/or (R) the "remedy" if that particular heat stress does result in down time or damage to your telescope.

P1   Leaving the telescope in the hot sun uncovered.

S1   Simply, don't do it. It is never a good idea. The

only solution to this (like setting up for a weekend star party where your scope must be left outdoors) is to cover the scope with a canopy, not a tarp. A tarp or even a sheet will still build up a significant amount of heat under it, particularly if tied at the bottom. Never cover your scope with plastic, as this traps moisture that can ruin the scope and its circuitry. Use a canopy instead, and then cover the telescope with a sheet beneath the canopy; since the canopy does not rest directly on the scope, air is allowed to circulate and continually cool in the shade of the suspended canopy.

P2   Leaving the telescope in the hot sun covered.

S2   See above ... never do this!

P3   Internal heat in the telescope OTA.

S3   A big problem taking out and setting up your scope early, particularly with the Maksutov design, is that a closed optical tube assembly (OTA) will build up the day's heat. This can cause two problems of significance: (1) long cool-down time to observing and (2) damage.

R3a   Use the canopy discussed in S1 above to keep the scope as cool as possible during the day.

R3b   Wait until dusk or slightly before to take the scope out into the evening air and allow at least one hour for the ETX-90 and two hours for the ETX-125 to reach thermal equilibrium.

R3c   You can use the "chimney trick" to accelerate cooling of the OTA. Do this carefully to prevent the accidental entering of debris or dust into the OTA. All you need to do is to orient your telescope so that the lens (front of the scope) is facing toward the ground. Remove the end cap of the ETX scopes until the rear of the scope is open; in the ETX you will need to flip the small mirror as you would to "view" straight through the scope (you will be able to see the secondary through the opening in the correct position). Place a clean lightweight cheese-cloth (or similar lace-like fabric) over this opening to block dust and debris, and then merely wait about one-half the time. Your scope will cool internally twice as fast, as the port acts like a chimney, rapidly dispelling warm air otherwise trapped inside the tube.

P4   Excessive heat damage to the OTA.

S4   Safeguard as in S1. However, likely damage that can occur is in the cements that are used within the OTA, particularly that holding the secondary baffle tube to the meniscus lens of the ETX telescopes. This thin metal baffle is held in place only by a thin ring of adhesive that becomes like jelly when the temperature gets excessive in the scope.

R4   If your baffle does become loose and begins to slip, you will know it. Before you can actually see this with the eye, you will notice flares from images of bright objects such as Mars and bright stars, flares that you did not have before. If you notice this, unfocus (turn your focus counterclockwise) and look at the out-of-focus image. You will see the "disk" pattern of the star; if it appears oblong rather than circular, then your baffle tube has slipped. It will be necessary to carefully unscrew the entire end cell that holds the meniscus lens from the OTA (it may take some work to get it moving at first, and be very careful not to damage the pretty blue tube). Once out, you can get a replacement adhesive ring from Meade or you can merely use a very good temperature-resistant glue (something that never actually "hardens" but remains flexible as the temperature changes) very sparingly against the flat edge of the baffle that mounts to the glass. Use the mirror itself (the secondary) as a circular template to place the baffle. Let the glue dry for about two hours in a protected environment, and then reattach via screwing the cell and its lens back onto the OTA firmly (no need to do any alignment whatsoever); and that's all there is to it!

P5   UV fading of telescope exterior.

S5   Even if kept indoors and covered, light from windows, even in the shade, can eventually bleach the beautiful colors of your ETX scopes. I highly recommend (in addition to keeping out of direct light) using Turtle Wax's *Scratch Gone* soft wax for new car finishes. This comes in many colors and that is the key to buying it; if it is Turtle Wax soft and it has a choice of colors, you are getting the right product. It has a UV block and is excellent for both the black plastic (and metal on the LX) and the blue tube. Use this very sparingly (just moisten your rag with it) and rub gently and evenly on all surfaces; then buff. It resists fingerprints and UV sunlight and makes your scope look like a million bucks!

P6   Telescope electronic circuitry (other than the Autostar).

S6   During tests of the Autostar on 90+ degree days in mid-April 2001, I also exposed the mounting of an ETX-125 (less tube assembly) to the same temperature/sun exposure as the Autostars. These conditions would closely simulate about a two-hour observing session of the sun with the sun at its highest point during the day. The temperature on both days reached at least 92 F in my observatory for a prolonged period of two hours of direct sunlight. Although some major problems arose with the Autostar, there were no adverse effects to the internal telescope circuitry or electronic function within the drive motors/ encoders whatsoever.

P7   Protecting your optics from direct sunlight.

S7   It is easy if you are readying for a star party to inadvertently allow the direct Sun's rays to enter your telescope assembly; you should always (for your safety and that of the scope) point the telescope in any direction away from the Sun. If outdoors, remember your finderscope; direct sunlight will melt the crosshairs within about 8 seconds flat, as the Sun might inadvertently pass across the field of that little scope.

P8   Accessories left in sunlight.

S8   Your eyepieces and all accessories should be kept in the shade. Never put your accessory case out in direct sunlight, not even for ten minutes. It is like a little greenhouse and will ruin your eyepieces, particularly any of them (and many do) that have cemented components. If you carry a flashlight in your accessory case, it too can be damaged by excessive heat.

P9   Transporting your telescope cargo in your vehicle in the summer.

S9   Never, never lock up the telescope in the trunk of your car, even while the car is moving at 70 mph down the freeway toward vacation. The inside of your trunk does not understand the chill factor of the speed, and sunlight bearing down on the metal trunk lid will eventually heat that interior up to 150 F. I always recommend, if possible, transporting the scope and accessories in the back seat and all nontemperature- sensitive parts (i.e., the tripod, wedge, charts) in the trunk.

Your Telescope in Summer Heat – Part 2: Autostar – extremely important – please read!

Just when we thought we were getting our Autostar away from the environmental perils of winter's cold, here comes summer's heat. And, boy if you thought you had "cold temperature problems" this past winter, please take notice here.

P10 Battery power and hot weather.
S10 Not a problem – unlike winter months, your internal batteries (or your external DC power station) actually will provide a somewhat enhanced output during summer, provided they do not get excessively hot.
P11 Prolonged exposure of the Autostar to direct sunlight for very short periods of time – malfunction!

Important note: The following descriptions are what prompted me to write this very important guideline to summer safety for the ETX telescopes. As I was training the motors on an ETX one afternoon, I began to run into some serious malfunctions related to the Autostar that at first I could not identify. After examination and further testing the next day in 92-degree heat with two Autostars, it was learned that direct exposure to sunlight is very detrimental to your Autostar and will result in at least the following two situations in less than ten minutes of exposure (Do not attempt to duplicate. Permanent damage to your Autostar will result!):

1. The Autostar display will no longer function, although the commands of the Autostar can still be entered. What you will get will be a display of filled "boxes" all the way across both lines of your display screen. This will not go away until the Autostar has cooled for about 30 minutes, but your problems are not over.

2. Not all commands entered into the Autostar once the display goes blank (1 above) result in proper functions being obeyed. Again, about 30 minutes is required before the Autostar cools to where some commands can be correctly accepted.

3. Speed settings on the #497 Autostar (the number keys "9" fastest, "1" slowest) become totally inoperative

and everything runs at the fastest speed when the Autostar becomes overheated. Letting the Autostar cool down as above does not always result in your speed function returning; and so,

4. A total reset (obviously requiring all new user data entry, owner information, site specifications, telescope type, and Train Motors) must be done to put the Autostar back into normal operational mode once it has become overheated by exposure to the Sun and allowed to cool down.

I used both of my "test Autostars" and was able to (thank goodness) successfully revive both after the alarming results of only 12 minutes in sunlight. It is important to note that this is not a "maybe this won't happen to me" situation; it will. I ran the tests throughout two successive days on a hot muggy April week here in Arkansas; and believe me, it will do it over and over again.

So, I go back to my 40-year-old rule that I have preached to all telescope users, even way before microprocessors took the place of Magnusson Setting Circles, before Maksutovs began outperforming my old Unitrons.

Never subject your telescope to any condition, permanent or temporary, that you would not subject your 6-month-old infant child to. They are both just as sensitive, and certainly both are just as temperamental and difficult to please.

Importance of Balance

Earlier in this book, I talked about adding accessories, perhaps lots of them, to the ETX. I touched upon the importance of maintaining balance to avoid tracking or tipping-over problems. In the following article, we learn more on the importance of balance (courtesy P. Clay Sherrod).

The Importance of Balance in the Fork-Mounted ETX Telescopes

Lately, with more and more accessories for the ETX-90 telescopes being offered in the marketplace, I am getting inquiries regarding the importance of balance of the telescope and its mounting for proper use and continued good service of the telescope system over time.

Also, I am beginning to see coming in for my Supercharge tune-ups many telescopes whose mountings are suffering badly from excessive wear and slop due to use in severely imbalanced conditions over time.

Indeed, improper balance of the fork-mounted ETX is crucial and definitely will adversely affect accuracy, GOTOs, and tracking, and likely result in random slewing conditions if balance is not maintained in both axes. In addition, it will greatly shorten the precision and life expectancy of your tracking system. This is particularly true, more difficult to achieve, and perhaps more crucial when your telescope is mounted in polar mode.

In alt/azimuth (alt/az), care must be taken for good balance (you want a little "load" on the front of the optical tube assembly to assist in maintaining minimum backlash of the mechanical aspects of the drive/slewing motors) but primarily in the altitude axis. In polar, balance should be achieved in both declination (altitude) and right ascension (azimuth) axes and is not easily done, as the center of gravity will shift on you in different parts of the sky.

This is particularly true when heavy accessories are added to the front or back of the telescope optical tube assembly, such as the electric focuser, a piggybacked camera or guide telescope, a heavy eyepiece (such as Meade's 14 mm UWA), or a camera body mounted at the prime focus of the telescope. It does not matter which model ETX you are using; balance is important and crucial to the life of the mechanical parts of your telescope.

In alt/azimuth mode you must simply unclamp the altitude ("up-down") axis prior to observing and check to see if the telescope tube swings unduly one way or another too freely; merely add an appropriate amount of weight to the opposite end (usually the front end if a camera or other heavy accessory is added) until the excessive weight seems to be compensated for.

In polar mode the fork mount balance point will change as your direction of pointing changes in the sky and hence a sliding counterbalance system is really necessary to compensate for this shifting center of gravity. If using polar, and more and more ETX users are doing just that, adjust your balance according to the position of the sky you are in; since you cannot unclamp your ETX once you have gone to a position (you will lose your alignment), you must do this through trial-and-error. I do this in the daytime and make notes of what is required in "southeast", "near-meridian", "northwest" and so on, sketching little diagrams of what weight is necessary where after a GOTO. Then I merely add on the weight while the drive is running after each GOTO, and the scope does fine.

You will notice from some trial-and-error balance testing that many times you may be well balanced in declination, yet your right ascension (your main driving axis) is severely out of balance. This can be particularly true when pointing the telescope far southeast or southwest with heavy accessories on the eyepiece end.

If you do not balance properly, not only is your accuracy and tracking affected, you will slowly wear out the small motors, add looseness to your gear trains, and cause excessive slop in your clutches. Take care of your ETX by assuring proper balance at all times and in all positions.

Seeing Conditions and Transparency

As mentioned previously, the Earth's atmosphere affects the quality of observations through a telescope. Sometimes objects are crisp and steady; other times they are blurry and seem to swim around in your eyepiece. In this article (courtesy P. Clay Sherrod), you will learn more about "seeing" and sky "transparency", how it affects observations (and astrophotography), and how to determine whether observing will be good or poor.

Seeing Conditions and Transparency: You be the Judge of the Night Sky!

Every time winter rolls in frosty nights our way, questions about telescope performance arise. Under these crisp, clear winter skies of North America, the stars almost reach out and grab us as they twinkle against the inky background of distant space.

"Why can't I make out Cassini's division at 75 power? I used to do it all the time. I'm worried about my telescope!"

"I looked at Jupiter and could see the four moons (Galilean satellites), but for some reason all I could see was two brown bands. It was all washed out. What's wrong with my ETX?"

"While looking at Betelgeuse last night, I could see the nice red color okay; but now I'm really sick. I think something's happened to my scope. Coming straight up (it was nearly overhead) I could see a bright flare of light, making the star look elongated. Should I send the telescope back?"

Do these sound familiar? And to make it worse, these actual concerns were sent to me from people who spent a night in the freezing cold to "enjoy" the wonderful hobby of amateur astronomy.

In every case, every telescope is just fine. It just so happens that from where they were observing, the sky was too clear. Those "twinkling stars?" They are to blame (or at least what you see of them). Read on.

"Too Clear?" you say. There's no such thing. Yep, there is for high-power viewing and maximum resolution. And it usually happens in winter.

The term "too clear" refers astronomically to really good "sky transparency," our ability to peer through our own atmosphere filled with water vapor, gases, pollution, dust and even light toward the dark of space. Without the atmosphere, every night would be perfect, even when the Moon was out!

The term "steady image", known as *seeing* astronomically, is totally unrelated and indeed can be thought of as inversely proportional to the transparency. Simply put, the better the transparency (like in winter months) the worse the seeing or image steadiness is typically.

Imagine it this way: transparency relates to how deep you can see in space through the obstacle of the earth's air. The better the transparency, the fainter the object or star that you might be able to see.

On the other hand, seeing has absolutely nothing to do with transparency; it is a gauge of how perfect, or steady, the image remains while you view it.

Now before going into detail, remember the "really-neat-rules":

1. If the night is perfectly transparent, it is a low-power night, but not necessarily a high-power night.

2. If the night is perfectly steady, it is a high-power night, but not necessarily a low-power night.

3. If the night is transparent and steady at the same time, it is a perfect night, both for low- and high-power viewing.

4. If the night is cloudy, it's Miller time.

Sky Transparency and what to Expect

In the old days when I was a professional astronomer, sky transparency was judged both visually and photoelectrically by a scale of 1 to 6.6, with 6.6 (the hypothetical limiting magnitude of the faintest star visible to the eye before light pollution) being perfectly deep clear. On the other hand 1 would represent **only** the brightest planets and stars (magnitude 1 or brighter) being visible. Such a case would be in the summer months when smog or fog "capped" the night air preventing all but the brightest objects from being seen. As we will see, such nights are the very best for "steadiness".

Here are two charts (Figures 6.2 and 6.3) that will allow you to judge your transparency from any location. The Hyades is excellent for fall, winter, and early spring; and the Little Dipper (Ursa Minor) is excellent for spring, summer, and fall, since it is circumpolar (mid-northern latitudes) and can be seen nearly all night. However, it should (like the Hyades cluster) be used as a guide only when the stars are reasonably high in the sky. In transparency, particularly for astronomy, there is a factor known as "atmospheric extinction" that causes starlight to diminish as the object decreases in altitude toward any horizon. It is caused by the "stuff" in our air and is particularly troublesome during high humidity and smoggy conditions.

As I mentioned, transparency is a measure of how clear the sky is; this will allow you to see fainter stars and deep sky objects but will not necessarily allow you to use more than about $25\times$ per inch aperture, and then only for deep sky objects. Typically (there are exceptions) the clearest nights are the unsteadiest.

This is because of meteorology, not astronomy. All day long, whether cloudy or raining or snowing, the Earth's crust and all upon it soak up the radiation from the Sun in wavelengths that we cannot even see. Objects on the ground will hold that energy so long as: (1) the

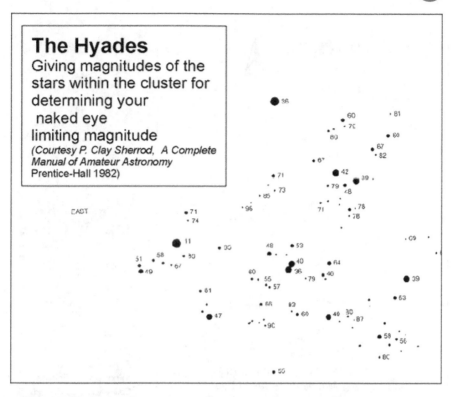

The Hyades
Giving magnitudes of the stars within the cluster for determining your naked eye limiting magnitude
(Courtesy P. Clay Sherrod, A Complete Manual of Amateur Astronomy Prentice-Hall 1982)

Figure 6.2. The Hyades star magnitudes. Courtesy P. Clay Sherrod.

air around them is near the same temperature as the heat-soaked objects or/and, (2) the object continues to receive an equal amount of radiation (heat) as it is losing into the air over the same period of time.

Okay, what happens at night? The same thing that happens in the desert on a hot summer day when your are driving down Route 66 at mid-afternoon: the ground is much warmer than the surrounding air, and heat (from the ground) always dissipates into cold (the air). When on the desert highway, we see this as a wavy image (mirage) as if water were standing on the road in the distance. At night, we see this rapidly rising air as twinkling stars.

This night-time convection is called attaining thermal equilibrium (I won't use terms like that anymore) and continues until the ground is the exact same temperature as the air.

This condition (happens every night to a degree) does not affect transparency, although it is the most detrimental factor involving seeing.

So, the first thing you will do each night when you go out is to judge the transparency by determining which

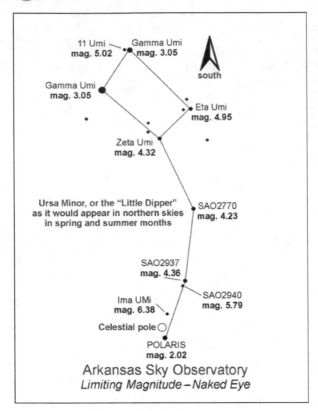

Figure 6.3. Ursa Minor star magnitudes. Courtesy P. Clay Sherrod.

of the stars from the charts you can distinguish. It is fun and interesting to keep track; for example, in Arkansas the month of October is by far the best combination of transparency and seeing of any month and also just happens to be the most weather-friendly month as well.

Transparency is usually worse in summer (inversion layers of air, stagnation of air) and best in winter (cold blasts of artic air bring pure, arctic-filtered and clean air southward).

Once you have checked out how clear the sky is, let's move on to see if you can make out the Great Red Spot or the polar region of Saturn with some really cool high power.

Atmospheric Seeing Conditions

We have seen that really clear skies don't normally indicate that the night is also a really good night for high-power planetary work or splitting that double star you've been promising to do but keep putting off. Also, we understand now that it is the Earth that is responsible ("... baaaad Earth!") for giving us nights when we think we should be able to see Cassini's division on Saturn ("... but I saw it last night!") but it is invisible except in fleeting glimpses.

Seeing can be rated on a scale of 1 to 5, with 1 good only for lowest power viewing and 5 perfect steadiness. Remember that I said that in summer stagnation and "inversion layers" cap our atmosphere to prevent really transparent nights? This is good for seeing, because the air below the inversion is trapped so the heat cannot move rapidly in these upward currents. Consequently, the air immediately above your telescope is very, very steady.

Early at night is the worst time for steadiness; just as soon as the Sun sinks low enough in the sky, the ground begins giving up the heat it has basked in all day long. Those heat currents rise rapidly upward into the sky. Let's look at how we see them:

Naked Eye

You can use the twinkling of stars to get a quick handle on the night's seeing conditions using an imaginary scale. Look toward your most unobscured horizon (where you can see stars closest to the distant horizon) and mentally divide the sky into four equal parts from that horizon to overhead. Each one of those parts we'll call a "Zone" for seeing evaluation. Zone 4 is from the horizon to 1/4 way up to the zenith (directly overhead); by comparison, Zone 1 is overhead. Now look at the brightest stars in each of your quadrants starting with the horizon; are stars in Zone 4 twinkling? How about Zone 3, a little above the first zone; are those stars also twinkling? Now Zone 2, nearly overhead.

Now it's easy: if only the stars in Zone 4 (lowest to the horizon) are twinkling rapidly then your seeing is at least a 4 on a perfect scale of 5; if, however, a bit higher up in Zone 3, the stars are also twinkling like those in Zone 4, your seeing is now 3; if the stars are twinkling rapidly all the way to the zenith (Zone 1), then your seeing is a pitiful 1, the worst it can be.

Remember, this has nothing to do with the transparency of the sky.

If, on the other hand, none of the stars twinkle all the way to the horizon, then your seeing is possibly a 5. To determine whether it is that good requires that you move from your naked eye to the telescope.

Telescopic Seeing

So far we have determined that you might have a really steady night of at least 4 and perhaps a perfect 5. Let's check. Turn your telescope to a very bright star that is close to overhead. Make sure to center it precisely in your field of view. Now check your air steadiness through these two steps:

1. Precisely focus the star until the image (the "Airy disk") is as small as possible; the smaller the telescope, the larger the center bright point, or Airy disk, will be. Once focused, you should begin to see clear diffraction rings surrounding the Airy disk that appear to be thin rings of light. Under steady conditions you will note that (a) the Airy disk does not move nor does it change size; and (b) the concentric rings remain fairly steady, both in motion and brightness. You should see perhaps three rings in most telescopes, perhaps more.

2. Now the tough test. Put the same star out-of-focus until you see a medium-sized disk (not too big) of light with a center dark portion, looking much like a donut and its hole in the center. Put this defocused star in the dead center of your FOV; well-collimated optics will show the dark spot exactly centered on the bright disk. If your star moves toward any edge of your FOV, the dark disk will shift in the opposite direction. So keep centered. Once defocused, look carefully at the bright disk. You will see alternating rings of dark and light, very, very fine. Examine all of them and see if they appear to move, or oscillate. If the image moves like a squirmy amoeba, then the seeing is terrible and you can forget high power. It's galaxy viewing time tonight. On the other hand, if the image is very steady and uniformly bright, the seeing can be excellent.

It's really simple to evaluate. Always remember: there are going to be nights when you want to give up and blame the telescope for poor performance when you might not have the contrast or detail you are used to seeing. Indeed, it is probably the seeing conditions and not your scope. If you find yourself in this situation, "... take that lemon and make lemonade!" as they say. Develop a mindset that "... tonight, I'm going to find those galaxies that I have been looking for!"

A Closing Word of Caution!

The best seeing and the deepest transparency cannot offset a telescope that is suddenly rushed out of the house and plunged into the cool (or cold) night air. Here is the rule of thumb that we have always used, no matter how big or small the telescope: put the scope in the observing location (in shade if the sun is still shining) for a total of 20 minutes per inch of aperture of telescope. A three-inch needs one hour to adjust (thermal equilibrium – heck, I did it again!), while a five-inch requires two hours. But the wait is worth it.

This is why professional observatories (even with forced ventilation systems) open the domes before sunset and commence through dusk to allow the scope to equalize with the night air. The longer the night goes on, the more "married" are the temperatures of your telescope and the air around it.

Now that you understand all of this (isn't it interesting that most of this has absolutely nothing to do with your telescope itself?), you probably have a better grasp of why the scope that you love so much sometimes can be as ornery as a pet cat, tuned to your beckoning call one minute, only to ignore your very existence the next.

Here's to the best in seeing for all of you.

Limiting Magnitude Charts

If you have read the previous article on seeing and transparency, you can begin to appreciate just how involved this hobby of amateur astronomy can be. Of course, you do not have to worry about such things; but if you do, your enjoyment will definitely increase and your frustrations will decrease. With this article you will learn something about telescope performance (courtesy P. Clay Sherrod).

The Arkansas Sky Observatory Telescope Limiting Magnitude Determination Chart

The Meade ETX telescopes are incredible instruments, both in electronic innovation and in their optical performance. I have published a field test on the ETX Web Site in which the results of some stringent parameters were tested with my ETX-125 5-inch Maksutov–Cassegrain. Many amateurs who have not had the opportunity to really use an ETX may not understand that the optical design of these telescopes is pretty close to perfect and their performance reveals just that.

I have had a lot of correspondence from ETX users and other astronomers regarding "limiting magnitude" when using the telescope visually. Visual limiting magnitude must be differentiated from photographic magnitude, in that the camera's film (and CCD imaging) has the ability to accumulate light, like a sponge slowly soaking up water.

In addition few amateur astronomers realize that there is a significant difference in their abilities to see certain color stars, normally with very reddish stars appearing more difficult to discern than blue or white ones when very faint.

The telescope's (or the eye's) ability to reach a limiting magnitude (say, 11.8 with the ETX-90 or 12.8 with the ETX-125) is termed the visual threshold, that point where the very faintest of stars sometimes can be seen and sometimes can't. If you can, indeed, get even a momentary glimpse of that faint star, you can honestly attest to the fact that the magnitude of that star is your limiting magnitude.

There is considerable discrepancy in the literature as to the limiting magnitude of any optical system, including the human eye. In a telescope it will be restricted by:

1. the type of telescope, i.e., refractor, reflector, catadioptric;
2. the type of glass the image must pass through (including eyepieces);
3. the transparency of the dark skies in which the scope is used;
4. the visual acuity of the observer; and,
5. the quality of the optics, including coatings.

There are formulae available in all the books that I will not bore you with; from those formulae I have prepared a mean value, an average of sorts, of all of them and offer the list below. My 32 years in astronomy have shown me that this list is, indeed, very close to actual performance.

Under the darkest conditions (see below):

Aperture (inches)	Limiting magnitude
Human eye	6.5
2.5	10.5
3.5	11.4
4.0	11.7
5.0	12.8
6.0	13.2
7.0	13.6
8.0	13.9

and so on. Conditions vary, scopes vary, and observers vary. And so will a limiting magnitude from scope to scope. But not by more than 0.4 magnitudes under the identical conditions at the same instant with two observers of equal visual acuity.

Now, that being said, let's find out what we can see.

First you must determine how visually acute you are. I noted above that on dark skies you might be able to glimpse a star magnitude 6.5. That's pushing it under perfect conditions and age is most definitely a factor. The Native Americans of North America used the Pleiades as an eye test as well as Alcor and Mizar in Ursa Major. In the latter, if the "rider" could be seen atop the "horse," the young man was deemed suitable to become a warrior if all other tests (no drug tests back then) proved likewise. Many times skilled observers can see to magnitude 7.1 with the naked eye on high desert dry mountains and from peaks such as Mauna Kea in Hawaii, high above the Earth's vapor layer.

We, too, can use the asterisms of the Hyades and Pleiades in the constellations of Taurus to determine our visual (eye) and the telescope's limiting magnitude.

The first chart (Figure 6.2, the Hyades star magnitudes) is a star diagram of the Hyades (from *A Complete Manual of Amateur Astronomy*, P. Clay Sherrod, 1982; permission by the author) giving the magnitudes of stars from bright Aldebaran to the very faintest. In addition to your eyes, try your binoculars on this chart to determine their limiting magnitude as well.

When using such charts, be sure to use a red flashlight to view or sketch on a copy of the chart to

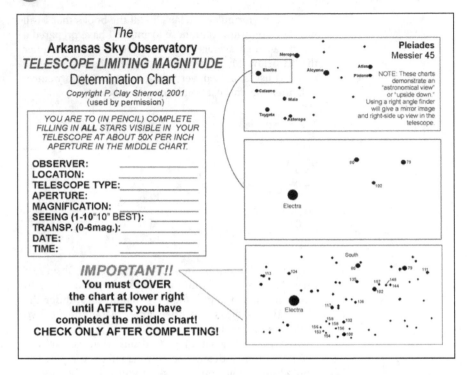

The
Arkansas Sky Observatory
TELESCOPE LIMITING MAGNITUDE
Determination Chart
*Copyright P. Clay Sherrod, 2001
(used by permission)*

*YOU ARE TO (IN PENCIL) COMPLETE
FILLING IN ALL STARS VISIBLE IN YOUR
TELESCOPE AT ABOUT 50X PER INCH
APERTURE IN THE MIDDLE CHART.*

OBSERVER: _____
LOCATION: _____
TELESCOPE TYPE: _____
APERTURE: _____
MAGNIFICATION: _____
SEEING (1-10"10" BEST): _____
TRANSP. (0-6mag.): _____
DATE: _____
TIME: _____

IMPORTANT!!
**You must COVER
the chart at lower right
until AFTER you have
completed the middle chart!
CHECK ONLY AFTER COMPLETING!**

Pleiades
Messier 45

NOTE: These charts
demonstrate an
"astronomical view"
or "upside down."
Using a right angle finder
will give a mirror image
and right-side up view in the
telescope.

Figure 6.4. Limiting magnitude chart. Courtesy P. Clay Sherrod.

maintain your night vision. Even when using the red light, you must wait about two to three minutes after illuminating the chart and then turning off the light to commence viewing to allow your eyes to adapt.

The second chart (Figure 6.4, Limiting magnitude chart) is a challenge. This is a detailed three-part chart of the Pleiades star cluster, centering on the star Electra. The first Pleiades chart shows the familiar pattern provided by the eight stars of the Seven Sisters. All of these should be visible to you on a dark night.

Like the naked eye test, only try this on the darkest of all nights and only when the clusters are nearly directly overhead.

The second Pleiades chart is your target: locate Electra and center it at high power (about 50× per inch aperture); identify the three stars that comprise the triangle just east of Electra; they should be visible in your telescope. This pattern, including and surrounding Electra, will be in your field of view at 150×.

Now, without looking at the lower right chart (the one that gives you what stars are really there and their magnitudes) sketch in with a pencil (on a copy of the chart) every star you see through your telescope; take your time; take some breaks. It should take at least one hour, and it's worth it.

When you're done, compare what *you* have drawn with what I have provided. The numbers given (the largest being the faintest stars) are without decimal points; if you have indicated a star of "135" and no larger number, that (13.5) is your limiting magnitude.

There is a distinct advantage in using the Pleiades in that nearly all the stars are "O" and "B" type whitish and bluish stars and more visually attainable. Also, it eliminates any error that could be introduced if you were, say, "red sensitive."

Important note! These charts were developed by me in 1976 for an asteroid identification program using large professional telescopes. Those telescopes always photographed and displayed "astronomically correct" images, that is they were upside down but correct right and left. Unfortunately, nearly all telescopes today utilize right-angle attachments that give an upright, but mirror-image view. Consequently, you may need to orient yourself to the star pattern surrounding Electra. It is very important that you identify the small triangle composed of the 7.9, 8.0, and 10.2 magnitude stars in relation to Electra. Another alternative would be to look at this chart in a mirror and turn it upside down. That would provide the correct orientation in some telescopes. A third and best alternative would be to use a "visual back" that accepts an eyepiece onto the rear cell of the telescope and view directly through the instrument, making your view exactly that of the charts.

If you have the terrestrial erect image prism for daytime viewing, you can use that and simply turn the chart upside down. However, this prism has optical elements that will reduce your limiting magnitude by as much as one-half magnitude.

. Good luck. Not only is this project fun, but it will educate you very well on the capabilities of your pride and joy – that neat little ETX telescope.

House Party Observations

On the ETX Site there is an area where actual user observation reports are posted. Also on the Site are some Observational Guides and References. Together, these can help new owners learn just what can be seen with the various ETX models. When combined with this book, you will be better equipped to match your expectations with the reality of visual observations through a small telescope, and you will be able to truly enjoy being an amateur astronomer.

One user observation report is mine on the "Sun, Jupiter, Moon, Mars during a House Party".

On 5 May 2001, we had a small house party. Several people arrived early (before sunset), and some stayed late (past midnight). I decided to set up several telescopes at different times to let people see some major items. For the daytime portion to view the Sun, I set up the Clear Night Products (home.earthlink.net/ ~barrycnp/) TeleDome Portable Observatory, put the ETX-90RA with a Thousand Oaks Solar Filter attached inside the observatory. Then for everyone's comfort, I put a Starbound Observing Chair inside the observatory. The TeleDome provided protection from the Sun's heat by shading the telescope and observer. During periods when no one was observing, I closed the observatory's side opening to further block the Sun. Several people were awed by seeing several small sunspots strewn across the solar disk.

After sunset I swapped the ETX-90RA for the ETX-125EC. Since the telescope was inside the TeleDome, I decided to fake the Autostar alignment as I planned to only show Jupiter using this telescope. I then slewed to Jupiter and tracking was perfect with the 2.2Ef version; Jupiter stayed centered in the 9.7 mm eyepiece (196×) for the hour or so that it was visible. Four moons were easily visible as were some cloud bands. Once, I parked the scope from the Autostar and when I powered on again, it began properly tracking right away. Again, many of our visitors were in awe of what they were seeing.

Just for grins I decided to unbox and set up my 40 year old Edmund Scientific 3-inch Newtonian reflector and let everyone see what I would have seen many decades ago. This telescope had been in its box for about 15 years; but following a finderscope alignment and collimation, it actually did quite well on Jupiter. It was only later when observing the Moon through it that I discovered what might be a problem; the Moon was not as bright as it should have been and appeared covered with a thin milky film. There was no distortion of the image, so I suspect that maybe the mirror coating has slightly deteriorated.

As the nearly Full Moon rose higher in the sky, I decided to set up the ETX-70AT and let everyone view the Moon. I used the 9 mm eyepiece (39×). At first while the sky was still fairly light, no filter was used; but as the sky darkened more, using the Scopetronix Moon Filter became mandatory for comfortable viewing.

What really impressed some visitors was the capability of the ETX-70AT system given its price. I pointed out that when it was purchased (in 1961), my old Edmund 3-inch cost $29.95; and the last time I saw a similar model in the Edmund catalog, it was $295. Now, for that same $295, you can get an ETX-70AT with tripod and Autostar GOTO computer. Quite a change from the limitations of the Edmund 3-inch. Some were even impressed when I told them the price of the ETX-125EC with Autostar and tripod; they were expecting at least a price of $2000. When I mentioned that they could buy a complete LX90 8-inch telescope system for less than that, they were totally amazed. I could just see the birthday/anniversary/Christmas planning going on in their heads!

As the party wound down, I kept wondering if Mars would clear the trees before everyone left. Finally it did rise high enough, and one person was still with us. So I took the ETX-90RA out to the front driveway (the only place where Mars was visible) and set up. Using the 9.7 mm eyepiece (128×), we were able to easily see some dark markings and I could see a polar ice cap. My wife and the visitor were not certain they saw the ice cap.

While this party was not planned as a star party, our visitors seemed to have had a good time looking through the different telescopes and seeing these major celestial objects. So, when you have friends over for a visit don't forget to let them experience the joys of looking through your telescope!

Autostar Objects Database

When the ETX-90EC and the optional Autostar GOTO computer controller were announced by Meade in January 1999, the ETX community began asking for more information about the Autostar than was documented in the manual. Users wanted to know how the Autostar worked, what was wrong when it didn't work as advertised, how it could be improved, and how to use it more effectively to make their observing sessions more productive and fun. In response I added the "Autostar Information" page to the ETX Site, which has grown tremendously in the past few years as more and more information was gleaned from the Autostar. The Site's resident Autostar expert, Richard Seymour of Seattle, Washington, USA, has been unceasing in his efforts to know all that is knowable about the guts of the Autostar and how its operation can be improved.

One of the typical questions is "besides our Solar System, what's in the Autostar database of astronomical objects?" After some research, the following article was developed (courtesy Richard Seymour) to answer that question.

List of Stars in the Autostar

Much of what follows is speculation, with some empirical evidence gleaned from using the Autostar to back it up.

Whenever you see a commercial product offering "9000 stars", suspect that they're using the Yale Bright Star Catalog.

The Autostar is probably using the 5th edition and frequently refers to entries in it as HR 5455 or some such. The "HR" is the Harvard Revision of the Bright Star Catalog. (Somehow, football-rival Yale's name got dropped.)

There is a NASA Site, at the Goddard Space Flight Center (GSFC), that has many, many catalogs, a number of which are combined and edited to create the Autostar's databases.

Visit:

ftp://adc.gsfc.nasa.gov/pub/adc/archives/catalogs/

and download the "key" file (as text).

It's the index to the hundreds of files at that Site.

The Yale BSC (above) is entry 5050. Hence it's under the 5/ subfolder, folder 5050. Each of the nine main folders also contains a key, but the top-level key is all of them combined.

The files whose names end in ".gz" need to either be fetched in binary mode or in clear-text by leaving off the ".gz" in your command string. The server will notice that and uncompress them before sending them. Beware that that will roughly triple their size.

Windows people can use a variety of programs to fetch the files. Your browser can do it; but like some of the ROM files from Meade, it might mangle the ".gz" form. If you are used to using "ftp", it will do it nicely; just remember to command "binary" transport. Macintosh people seem to universally use "fetch" as their favorite ftp program.

The 9110-star Bright Star Catalog (BSC, or HR) is available at

ftp://adc.gsfc.nasa.gov/pub/adc/archives/catalogs/5/5050/

The three files are:

1. readme (which describes the layout of the files);
2. catalog.dat.gz (the bulk of the file);
3. notes.dat.gz (the textual footnotes, many of which the Autostar displays as the scrolling message under many entries). (Compare the BSC notes on Polaris (BSC/HR 424) to what scrolls on the Autostar.)

That file also contains the SAO catalog numbers for the stars. Using that, I've noticed that the Autostar does not have the entire BSC. The Multiple and Variable star menus might be simply derived from keys in the BSC. Which stars are blessed to appear in the "Named Star" list versus which stars it knows names for escapes me. The BSC puts "common" names into the notes data. And Polaris has 11 names!

The GSFC Web/ftp Site is divided into nine major subsets. Category 5: Combined and Derived Data seems to be the most likely source of much of the Autostar data.

Other than the obvious BSC, it's anyone's guess (outside of Meade) which other catalogs truly comprise the rest of the bunch. The Nearby Stars might be from

5032A Stars within 25 pc of the Sun (Woolley+ 1970) or
5101 Nearest stars until 10pc (Zakhozhaj, 1979–1996)

or simply be ones from the BSC with distances below a certain limit. ("pc" is parsecs, and a parsec is about 3.26 light years.)

I suspect the Deep Sky comes from pieces of Category 7: Nonstellar and Extended Objects.

Entry 7118, for example, appears to be the New General Catalog (NGC) 7118. NGC 2000.0 (Sky Publishing 1988, ed. Sinnott), with 7001B Rev New Gen Cat of Nonstellar Objects (RNGC; Sulentic, Tifft 1973) sounding like a revised version.

I went bananas when I found this Site and haven't had anywhere near enough time to poke through all the files (and me with a 28k modem!)

Some specialized pieces of the Autostar database have probably been hand-prepared by Meade: things like the "Stars with Planets" list. The GSFC database doesn't contain a specific Messier catalog. (Since Messier objects span the gamut from star groupings (Pleiades) through nebulae, they likewise would be scattered across many of the GSFC files). Some of the scrolling comments involving "small telescopes" are possibly Meade's, too. Meade also have the data behind the LX200's controller and their Epoch 2000 software to draw upon. And where they get "3C273" as being a DeepSky>NamedObject is beyond me. (They're welcome to include it. It's a 12.8 magnitude quasar, but the scrolling text doesn't mention that it's the brightest one in the sky.)

Let's look at the Autostar version 2.x advertised database: 30 233 objects.

> 13 235 Deep Sky: the complete Messier, Caldwell, IC and NGC.
> 16 888 Stars sorted by name and SAO numbers

Well, I think that the SAO is about 240 000 stars, so there's been a bit of trimming. (I think the SAO is entry 5015 at the GSFC Site.) The IC (Index Catalog) is two supplements to the NGC. The original NGC was compiled in 1888, and the IC supplements were in 1895 and 1908. We didn't know about galaxies being what they are until well after 1900, when photographic emulsions became good enough to capture the faint light allowing us to finally see the structure of "island universes" such as the Andromeda Galaxy. And then it became clear that they weren't just fuzzy stuff in a randomly distributed sea of stars but were reflections of what our local star cloud (the Milky Way) must look like.

But that's a digression for another day.

Autostar Alignment and High-Precision Use

As a new Autostar (or any GOTO computer) user, the first time you go outside to set up that new telescope system, if you are not familiar with the night sky, you may wonder just which bright star is the correct one, that is, the one that the Autostar selected when doing an Easy Align. The same can apply when using the High Precision mode of the Autostar. If you, no matter what your experience level is, pick a different star from the one the Autostar intended, the alignment and GOTO will be incorrect. The following article (courtesy P. Clay Sherrod) should help with this star identification problem.

Seasonal Star Name Index for ETX Alignment, High-Precision, Star Directory Charts

Here are eight charts (Figures 6.5 through 6.12) that are at your disposal for use with your ETX telescope during observing.

Even a seasoned observer can become confused with the many sound-alike and similar star names associated with our brightest stars. Yet we must make use of these

Figure 6.5. ETX fall stars. Courtesy P. Clay Sherrod.

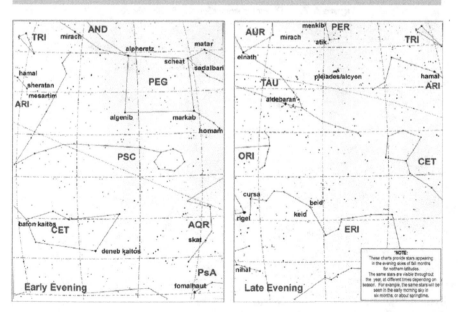

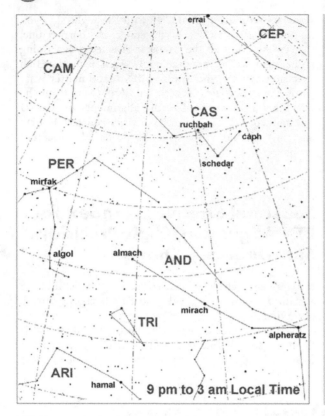

9 pm to 3 am Local Time

Figure 6.6. ETX fall stars 2. Courtesy P. Clay Sherrod.

Figure 6.7. ETX winter stars. Courtesy P. Clay Sherrod.

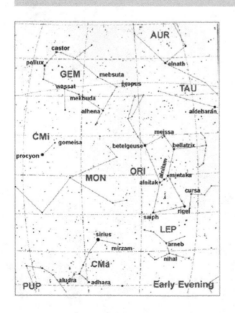

Early Evening

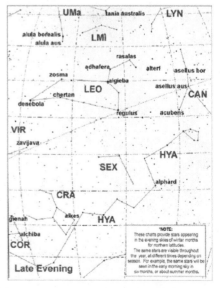

Late Evening

*NOTE:
These charts provide stars appearing in the evening skies of winter months for northern latitudes.
The same stars are visible throughout the year, at different times depending on season. For example, the same stars will be seen in the early morning sky in six months, or about summer months.

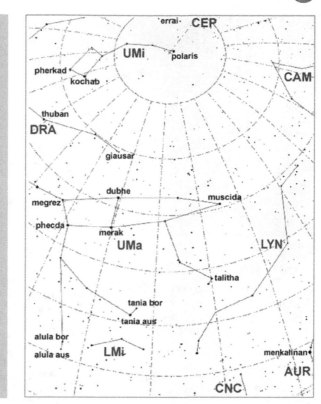

Figure 6.8 ETX winter stars 2. Courtesy P. Clay Sherrod.

Figure 6.9. ETX spring stars. Courtesy P. Clay Sherrod.

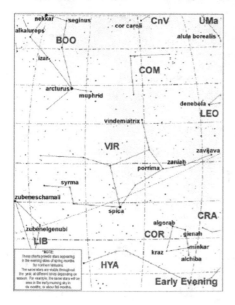

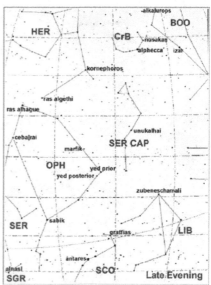

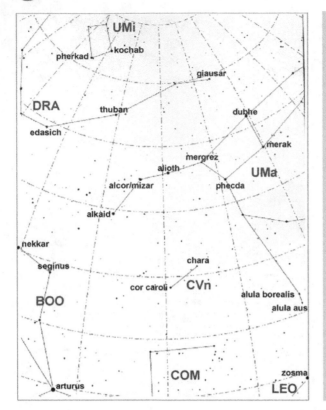

Figure 6.10. ETX spring stars 2. Courtesy P. Clay Sherrod.

Figure 6.11. ETX summer stars. Courtesy P. Clay Sherrod.

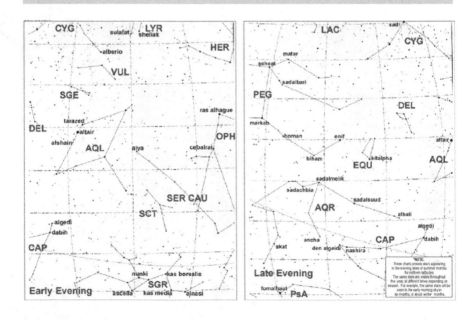

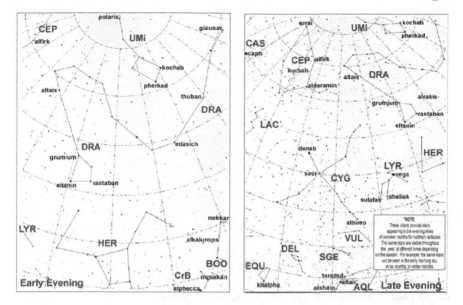

Figure 6.12. ETX summer stars 2. Courtesy P. Clay Sherrod.

stars to guide us through the sky, particularly in telescopic use. Don't be fooled for a minute about your knowledge of the sky and the star names within it.

With the new computerized GOTO telescopes that practically observe for you now, it would seem that our knowledge of the names of naked eye stars might not be as important as it was when, say, setting circles were in common use or when we did not even use setting circles.

But nothing is further from the truth. With your computer-automated tracking system, you perhaps need the backup knowledge of constellations, stars, and star names more now than you did before. Consider the following:

You are aligning your telescope in alt/azimuth for tonight's computer tour of the sky. Once your scope is in Home Position, you select "Easy" two-star align and press "Enter." It goes to the first star, say Capella, which you access and center easily and rapidly, then hit "Enter" again, at which time it slews the telescope for you to align on ... *drat*! It went to Deneb, and Deneb is behind the neighbor's house! So you hit the Scroll key and the computer selects another star for the second alignment and goes to it. Only this time the telescope moved all the way to Orion! Looking at the Autostar display, it is asking you to "Center Alnilam". Alnilam? Yes, you sort-of remember the star name, and you know it is somewhere in Orion. But then you look up with your naked eye. Orion is filled with bright stars and all you know for sure is Betelgeuse and Rigel. Look through your tiny finder and – again – there are at least six stars that could be Alnilam.

Which one is it?

If you guess, and you choose the *wrong* star of those six and press enter, your alignment will be off by that star's angular measure from Alnilam all night long. And you'll wonder what's wrong.

You, like me, like to find objects and appreciate the more accurate GOTO and tracking via "High Precision", whereby the telescope moves to the closest bright star to your desired object. The computer, thanks to the newest Autostar versions from Meade, actually tells you which star (by name) to look for. You look in the finderscope again, and several bright stars are there and you center the brightest. What if that star is not the one? Say you were searching for a new comet near the constellation of Corvus in the southern sky at High Precision. Likely, the computer will move you to either Gienah or Alchiba in Corvus and tell you which one it has chosen. Do you know which is which? Bet you don't! If it asked you to center Alchiba and you were confused (or simply didn't know) and instead centered Gienah, and pressed "Enter," the telescope will miss the comet by almost five degrees. Unfortunately, if you do not stop observing and re-align to north from scratch, you will be off that amount with every subsequent GOTO for the rest of the night. Most people do not even know what has happened, and quickly blame the Autostar or servos.

This introduction is merely to let you know that this has happened to me many times since I have owned my ETX-125. It will throw off your entire night's observing if you mistakenly align or High Precision to the wrong star.

The eight charts present all of Meade's Autostar bright stars in the star directory of the Autostar for northern mid-and-high latitudes (as of this version).

Note that there are four seasons: fall, winter, spring, and summer. Each season has two charts: the first for each season contains two star charts for stars seen in northern latitudes – one for early evening and another for later that night or the following morning. There is also a star chart (two for summer) for high declination and circumpolar stars that northern observers should have quick reference to. So if you want to cover a spring sky, you will need to use both ETX spring star charts (Figures 6.9 and 6.10) for a complete set of star charts giving proper names of all the reference stars that Autostar can throw at you.

Keep in mind that the stars names given on the charts are directly from the Meade star directory in Autostar and hence may not be the name you might be accustomed to; most are Arabic on these charts. Nonetheless, regardless of what they are called, you will never again have to remember which of the three "belt stars" in Orion is Alnitak or Alnilam or Mintaka. Just have a copy of the chart handy, packed in with your flashlight, eyepieces, and other accessories.

It was necessary to utilize the Autostar entries so that they would correspond with the star directory and, hence, (1) the alignment star selection, (2) the High Precision selection, and (3) the star reference listing under "Objects."

If in doubt about your alternate star or your High Precision star, consult the charts. They are easy to read and not cluttered with things you do not need for this purpose. The charts themselves have been adapted and considerably modified from the very early version (1995) of Star Navigator II from Meade. All of the named stars I took from the Autostar index and assigned to the appropriate star on the chart. All-in-all, wherever the Autostar sends your telescope looking, you'll be ready.

You'll be surprised how often you will use these charts. No more fumbling around with a flashlight attempting to open a field guide to the right page to verify your star; just look at one flat chart.

Now that you have access to so many stars so easily, why not start using "High Precision?" It locks onto a reference star (which you will verify and center using your star charts), and then it very accurately positions the scope nearly dead-center with the object you are searching for. I have found that if I really align my scope very carefully at the night's beginning, my ETX-125 will show the object in the field of view using my 26 mm Plossl (73×). That's "high precision." Try it. Now you can with confidence.

Enjoy some incredible accuracy and confidence in the sky by being a "master" of star names. It really makes your pursuit more enjoyable and hassle-free.

The best in seeing and the finest in GOTO.

Tracking Satellites

Many users enjoy having the Autostar track a satellite as it passes across their night sky. Most satellites appear as just a moving spot of light in the sky due to their small size. However, the International Space Station (ISS) is very bright and may even show a cross-shape in a moderate power eyepiece due to its large size. With the Autostar doing all the work of locating and locking on to the ISS, you can just observe and see what details you can make out.

Before you can use the Autostar to track a satellite, you have to give it some information about the satellite so it knows when and where to look for the satellite and in what direction it will be moving. Using this information, your Autostar will move the ETX to view a portion of the sky and wait for the satellite to come into view, and then it will start tracking the satellite's movement. But how do you get that information and get it into the Autostar? This is the purpose of the following article (courtesy Richard Seymour), which details the process for the ISS and includes an expanded explanation of the process using another satellite.

Tracking the International Space Station

Visit www.heavens-above.com.

Here you've got two ways to go (I use both):

The non-Autostar way is to tell it where you live and what you want to see, and it will draw star charts of the pass over your site.

Cool.

Cooler is: on that pass map's page, in the upper right corner, is a link to "orbit". Click on it. It will take you to a page that (scroll down past the path-maps) has the numbers you can give to the Autostar so it will track the ISS for you. It's fun! It works! (somewhat)

On the Autostar, go to OBJECT>SATELLITE>SELECT [ENTER]. Now use the scroll keys to see if the ISS is there. If it's not, then [MODE] back up until you're seeing

SATELLITE
SELECT

Now press scroll-down, and you'll see:

ADD [ENTER]

Name: ISS [ENTER]
Then feed in the latest numbers you can get (orbits change). After you've entered the ISS once, on future visits to this portion of the Autostar you can use the Edit function (scroll up from SELECT).

To track a satellite, press [Enter] when you see its name on the Select list. It will calculate the next pass within six hours. When it shows you a "Sat Rises at XX:XX", you can scroll down through other pieces of pass information. When you press [GOTO], the ETX will first swing to the LOS (Loss of Satellite), then it will swing to where the AOS (Acquisition of Satellite) will occur and start counting down the seconds until the expected passage. Press [ENTER] when you see the satellite cross the cross-hairs.

The "?" key on the Autostar explains this fairly well.

ISS TLE Interpretation

On the Heavens-Above page showing an ISS time, click on the "ISS" itself. That brings you to a page describing the satellite. Then, on the upper right of that page, click on the word "orbit".

Here is the link itself:

www.heavens-above.com/orbitdisplay.asp?satid=25544

Under the pictures is the two-line element listing. Below that are the values as a list. Here are some actual values:

1 25544U 98067A 01050.20978583 .00032487 00000-0
36862-3 0 6650
2 25544 51.5750 323.9431 0009808 113.8873 50.0677
15.61998931128644

The 01050.209785 becomes Epoch Year 2001 and Epoch Day 50.209785.

Move down to the second line. The 51.5750 is the inclination. The 323.9431 is the "RA Asc, Node". The 0009808 is the eccentricity, missing the leading "0", so you would key that in as "0.0009808". The remaining numbers on that line are in the order the Autostar wants them, namely, Arg. of Perigee, Mean Anomaly, and Mean Motion, which is expressed in Orbits Per Day.

The only tricky part is rounding the numbers that won't fit in the space the Autostar offers, such as Mean Motion: 15.6199893112864. When you get to the last spot on the Autostar's screen, just round up that digit if the following digit (which won't fit) is 5 or greater. Such round-ups may occasionally ripple back up (so 15.619989 might round to 15.62000).

More on Satellite Tracking

On a quiet day in the dim past, I visited www.heavens-above.com and told it where I live. It produces a page with links to tonight's satellite passes, cut off at various brightness levels. **Stop:** Bookmark this page. I choose the < 4.0 (less, i.e. brighter, than 4.oh) link. Click and go there. You're now facing a page of tonight's passes. **Stop.** Bookmark this page, too. This is the first page you should visit every evening. From tonight's list I choose satellites separated by at least five minutes. It is well nigh impossible (i.e. even more frustrating) to try to pack more in. I choose a mix of brightnesses, knowing (from practice) that satellites dimmer than 3.5 are difficult to see in my finderscope under my sky conditions. They're also usually far dimmer as they're first coming up, too. I choose ones that will have pleasing or interesting pass altitudes. Ones that pass directly overhead are difficult; the ETX has to spin 180 degrees in azimuth as the satellite goes merrily past. With practice you'll be able to manually catch up after that and continue tracking down. But they're worthwhile for honing techniques of starting the pass and finding them at all.

Then I visit each pass's chart, shuffle the Big Square view to show me the area where I'll first be seeing the satellite rise. Previous to Autostar version 2.2, I would center on about 25 degrees elevation, now I'll bring the horizon into the bottom of the frame since that'll cover the range of AOS points the Autostar offers. I'll either take good notes or print that page. Then I click on the [ORBIT] link at the top right of that page. That brings me to the TLE-bearing page. I tend to highlight the two-line parameter set (below the globes, above the labeled numbers).

Here's the set for a Lacrosse 3 pass:

```
1 25017U 97064A 01078.08554567 0.00000900 00000-0
15504-3 0 07
2 25017 57.0100 89.2234 0007000 158.2663 201.7337
14.68206719 03
```

These I copy/paste into a Notepad (Mac: TypeText, BBEdit) edit page, which I use to accumulate the sets I'll want tonight. On good days, I do this in the afternoon or the night before ("Next PM" link on tonight's passes page).

Now let's key a set into the Autostar. From the above, the 01078.08554567 (top line, near the middle) breaks into Epoch Year: 2001 (the "01") and Epoch Day: 078.08554567. The Autostar only accepts 078.0855; since the next digit is below 5, I don't round it up.

Now drop to the second line. The first item of interest here is the 57.0100. That's the inclination and will be the next thing the Autostar wants. The rest of the line: 89.2234 0007000 158.2663 201.7337 14.68206719 are in the order the Autostar wants them. Those are, respectively: RA Asc. Node, Eccentricity, Arg. of Perigee, Mean Anomaly, and Mean Motion (often listed as Orbital Motion).

Two notes: the eccentricity 0007000 gets a leading "0." when keyed into the Autostar, so it becomes 0.0007. Second note: the Mean Motion (orbits per day) 14.68206719 does round up to 14.6821, since the next digit is a 6.

On really well prepared nights, I might pretest a pass: fire up the scope and fake the time to a few minutes before the prediction. The Heavens-Above Site shows the 10-degree altitude point. Older versions of the Autostar software will pick AOSs in the 15 to 38 degree altitude range, newer versions start as low as 2 degrees above the horizon. If you don't like an Autostar AOS, just [MODE] and reselect the same satellite. It may take two or three spins of the dice, but it'll (randomly?) choose other altitudes along the same path. When you're happy with one, press [GOTO] and the ETX will first go to the LOS (Loss of Satellite) point, then swing back to the AOS point. It does that to avoid hitting the hard stops during the pass. In the heat of the actual pass, you can abort the "to-LOS" slew after it starts by a short tap on the [MODE] key; the ETX will then go to the AOS point. The Autostar starts counting down to the predicted AOS time.

I now press the 7 key to preset the manual slew speed.

On really good nights, I'll remember to bring along binoculars. Failing that (usually) I'll just glue my eye to the finder as we get within 60 seconds of AOS. No need to watch the Autostar; it'll beep as we pass zero seconds. If I see the satellite in the finder prior to the AOS beep, I'll judge if it'll hit the center or if I need to do a preslew fudge. If so, I'll do that before pressing enter. When the satellite hits the cross-hairs (yeah, sure, it does!), I press [ENTER] and the game's afoot. Assuming hit-the-center, I switch to the eyepiece, ready to "help" the tracking with short burps (or long pans) of the slew keys. I do not try to keep it centered; I let it drift around in the eyepiece and burp to overcorrect so it'll drift past again.

This really gets the old "which way do I press to move the satellite into view?" reflexes trained.

As the stars stream by in the background, I make mental notes of "oh, that's pretty!" for visiting later or perhaps even abort the satellite pass to visit now. Priorities, you know.

Well, that's it. Keep burping along until it gets irretrievably off-aim, or until it drops below the LOS point (at which point you're the tracking engine) or drops behind something opaque. (I've got to remove that chimney.) Catching up with over-the-top passes also develops aiming techniques that improve normal star watching. Then consult your "tonight's" list, and set up for the next one. Continue until exhausted.

I think it took six evenings before I even got a satellite into my eyepiece for five seconds. Now I can get four to six satellites held within my 13 mm eyepiece in an evening. But I sweat a lot doing it. (That's a very recent accomplishment. Practice, not the sweating.)

But wait! There's more! (or: Swiss cheese memory strikes again).

A footnote to all of the above:

When looking at the Heavens-Above TLE page, note the satellite's altitude range. If it's above 400 km or so, the orbital elements will "last longer" than something low, or frequently adjusted, like the ISS. I have had a relatively successful pass with two-month-old TLEs for a 580/600 km satellite. Rockets (the leftover upper stages of the launch systems) also tend to have longer lasting TLEs, since they're not adjusted for orbital position, whereas many satellites are frequently nudged to keep on station (or to go somewhere else).

Now fixed in v2.2Er, there were two bugs in v2.1Ek that would fight you when satellite tracking:

1. If you press [ENTER] to "pause" during a track, the system will not resume properly thereafter. Rather than pick up with the speed and direction changes appropriate for the spot in the track, they restart the motion list from the beginning. Yes, I've told Meade.

2. If you have an electric focuser, do not press the "Zero" key to engage it during a track. The main alt/az axes stand a very good chance of taking off in a rapid slew.

All that said, it really isn't as hard as it sounds. I've ripped myself from the computer screen, dashed outside, plopped down the scope, and been tracking a satellite eight minutes after powering up the Autostar. It can be done in less than three minutes (more frantic, more sweat).

Subtle secrets that help rapid setup (this is how I stargaze, too):

- Have a level, flat surface available.
- Have a due-north landmark that the "home" position can see.
- Park the scope at the end of the previous power up.

Why?

The level, flat surface (and a tripod provides one) is always needed. The landmark allows point-with-finder-before-power-up alignment to within a degree or two of due north. This matches the "park" status. Parking skips the Alignment step on the next power-up. The scope will power-up aligned as per the previous usage, corrected for current time and date. For satellites, that's quite adequate.

Super secret: Even if you're at a new location, simply power up the scope, answer the time and date, and press [MODE] to escape the Align step. Tell it the new location.

Fake an Align (i.e. just hit [ENTER] when it asks). Park. The next time you power up, the alignment will match whatever the scope would do if its initial guesses were perfect. And that's probably good enough for satellites. (Pretty good for off-the-cuff observing, too.)

Autostar External Power Source

The Autostar draws its power from the ETX; so wherever you want to use the Autostar, you need to have the ETX nearby. When observing, this is perfectly understandable. But what about those times when you are using a desktop computer to update the Autostar ROMs or to manually enter object data? Wouldn't it be convenient to have an external power supply that would just power the Autostar? If you are handy with electrical components and cable making, you can cobble together such an external power supply using the information in this article (courtesy Richard Seymour). Of course, you can damage the Autostar, yourself, or your home, so **do not** try this unless you absolutely know what you are doing and are willing to take the risks.

External Power for Autostar

Yes, you can power a #494, #495, or #497 Autostar without a computer.

You need to make a cable to replace the HBX cable (#495, #497) or to connect to the fixed #494 cable. You need to feed a nominal 12 V DC, but for stand-alone operation of the Autostar, anything from 9 V to 15 V is acceptable.

If you look at the bottom of the Autostar, the positive side of the power supply goes on the wire next to the RS232 jack.

Here's a picture, looking at the bottom end of the Autostar.

Ports in the Autostar (only the leftmost diagram for the #494):

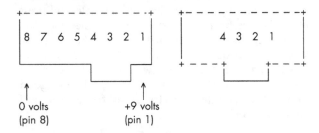

All the standard warnings apply: you can totally destroy your Autostar. There are no fuses or protective devices. Reversing the polarity or feeding power (or ground!) into any of the other six pins will (not "may") damage your Autostar.

Good luck.

Full pin labels

HBX Port in Autostar
(or #494 cable)

1: +9 V	5: A/A Data (ties to 7)
2: Aux Clk	6: Alt Clk
3: Aux Data	7: A/A Data (ties to 5)
4: Az Clk	8: Ground

RS-232 Port in Autostar

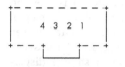

1 - Astar Rcv: data to Autostar (PC pin 3) (25/2)
2 - Astar Xmt: data from Autostar (PC pin 2) (25/3)
3 - not connected
4 - ground reference (PC pin 5) (25/7)

Autostar and Drive Motor Failure

Occasionally, the ETX and/or Autostar act up; the Autostar may display a cryptic message about "Motor Failure" or the ETX just doesn't respond to inputs from the Autostar. This article (courtesy Richard Seymour) provides some troubleshooting information that might be useful in some situations.

Motor Unit Failure on Power on

This article is in response to a situation where the Autostar displays a "Motor Failure" error message right after displaying the Sun warning. Previous to this the AC power supply had been interrupted.

Before trying a download (which will require the "hold-the-ENTER-and-SCROLLDOWN-keys on power up" override, known as "SAFE LOAD"), try these steps.

Plug in the default handcontroller. When you power up, it sits there blinking all of its lights. Tap any slew key. It will try to move both the az and alt axis a few degrees, clockwise and up (it's performing a calibration step). If it can and it gets good feedback, it will simply stop and wait for you with (most likely) the top light illuminated. If that's what you see, then the scope electronics and motors are probably happy.

If it encounters problems, it will blink all of its lights and may start repeatedly trying the calibration moves. Exactly what symptom you'll get varies with whatever failure you're experiencing. Mine started circling clockwise, with no alt motion.

If the ETX passes the handcontroller test, suspect the Autostar.

Next: have you suppressed the "Getting Started" message on the Autostar? The error message you're getting is probably from when the Autostar is attempting to perform a "deep reset" motor calibration step. If the Autostar has never done one, it inserts it between the "Sun Warning" and the "Getting Started" messages.

If it is attempting a calibration, it should display "Testing Motors" on the display (perhaps very briefly).

It sends commands to move the az clockwise, and then the alt up, and then checks the values it gets back. If either test returns "255" (or minus one, depending upon interpretation), it declares a motor fault. That "255" can arise from the motor circuits or a failure to communicate.

In v2.0h, it also locks that "bad" value into the "send next time" memory location, so subsequent power-ups may also have problems. I'm going to dig into that further; it might be a bug.

During normal power up, the Autostar sends the calibration information to the ETX drive base. There's no motion; it's merely sending the values it attained back when it did a calibration.

After a full download across a version jump (i.e. v2.0 to v2.1), the locations in the Autostar's memory, where things are stored during power-off, changes. So when you install a new update, you do have to perform a RESET (or the Updater should do it for you) to guarantee that things get placed in the new locations. Wild rides are the consequence of not performing at least one RESET after downloading a new Autostar version.

There are two forms of RESET: the one you can command from your Autostar keypad and the one that the Updater application can command. The Updater can command a deeper RESET than the keypad. This "deep" reset starts with a 1-Jan-0000 (or 1990?) date and automatically does a calibration.

When my ETX fried, the Autostar didn't declare a fault until after setting the date and time. Only the alignment step finally noticed that the beast wasn't happy.

Now the caveats: I suggest opening up the base and looking closely for mechanical damage: the yanked power cord could have hurt the little circuit card with the power plug (and RJ45 jacks) on it. If I'd done that before trying my standard handcontroller, I wouldn't have damaged it, too. (My problem was that the RA clamp in the ETX base had cut through the alt wires, shorting lots of things together.)

If you're handy with wiring, you could create a power-only cable for the Autostar and test it without the scope present at all. All it takes is two wires into the RJ45 jack on the Autostar. +9 V (or higher) goes to the edge of the connector closest to the RS232 jack; ground is the other end of the connector. That's pins 1 and 8 getting power, all others unconnected. Be careful, check your wiring and the final product with a voltmeter before attaching; and you can run your Autostar without any motors to cause faults.

Good luck.

Improving the Performance of your ETX

If you have read this chapter sequentially from the beginning, you may have noticed that we have progressed from easier (but still very useful) topics to more technical ones. For many users, those are all they will need to know and very likely not all of that. But for others who want to fix a problem with their ETX or Autostar or improve its performance, the following four articles (courtesy P. Clay Sherrod) will be invaluable. Certainly not all ETX telescopes and Autostars experience the problems discussed (and Meade continually improve both the ETX and the Autostar software), but the tips may still be handy to know. For more technical articles, visit the "Telescope Tech Tips" page on the ETX Site.

ETX-90RA "Sticky" Right Ascension Movement Fix

The original model ETX (known as the ETX-90RA now) could suffer from a sticky right ascension (or azimuth) movement. The problem is in the locking clutch that the RA lock lever is attached to. It has developed a "high point" somewhere on the small disk that pushes against the drive gear when the clamp is tightened.

I have a suggestion that will likely help without the owner either having to get inside the telescope or having to send it back to Meade. On the ETX-90RA, doing the following can smooth out most clutch irregularities.

1. First unclamp the RA lock to be as loose as it can be; and then rotate the assembly 180 degrees in one direction and then the other, about three times both directions.

2. Very gently tighten the locking lever just a bit until you can feel a tiny bit more resistance when you now attempt to rotate as before; with the clamping lever engaged in such a position, repeat the rotation noted in (1) above.

3. Now tighten the RA lock until the RA axis will move with some degree of pressure from you in either direction; again, rotate 180 degrees three times in both directions.

4. Now unclamp to the position in step (2) above (slight pressure) and turn the Slow Motion knob until you reach those points where you are experiencing resistance; stop there.

5. Now firmly clamp down the locking lever until the scope is tight in RA but you can still rotate the assembly by hand and override the clutch; work the axis back and forth through one hour of right ascension (15 degrees) by hand, not by the slow motion knob.

6. Move the scope back to the rough spots and lock onto them dead center and clamp even a bit more; now turn the knob half a turn clockwise and then counterclockwise several times.

7. Do the same steps at every point you feel this rubbing effect.

Over time this will reseat the clutch relative to the drive gear.

Good luck!

Performance Enhancement – Creating the Perfect "GOTO" ETX

Part 1 – Mechanical Considerations and Adjustments

This is the first in a multipart series of articles as a practical guide to understanding your ETX-EC, either the 90 mm or 125 mm telescopes. Essentially, this is a comprehensive step-by-step guide to enhancing the overall performance and reducing the many reported problems of these telescopes. This series, if you carefully read the instructions, corrects what you can correct (or return to Meade to fix under warranty) mechanically. If you correctly engage in downloads of new versions of the Autostar Updater and its supporting software and you properly initialize and align your telescope prior to its first use after doing the above, you will experience what I call the "Perfect GOTO ETX."

Those of us who have had our ETX scopes awhile realize the many frustrations that seem to be packaged along with our beautiful blue telescopes. We have problems that have been discussed on the ETX Web Site. Many times it has been all too convenient to blame the telescope when indeed it turns out that many, if not most, of the problems are introduced by a combination of user errors and incomplete and misleading communications from Meade (i.e., instruction manuals and support on really technical issues via Customer Service).

Overview of Major ETX Complaints and Problems

I have spent a lot of time experimenting, polling other ETX users, and examining the many quirks of the ETX GOTO Autostar system. They include:

1. Rubber Banding: This is a common problem that seemed to arise with the introduction of Autostar software v 2.1Ek using the A2.4 AutoLoader from

Meade (subsequent versions appear to have resolved the rubber banding problem). It is the phenomenon whereby you GOTO an object, center it carefully, and suddenly Autostar decides it does not like where you put it and proceeds to reposition the object where it placed it originally at about 3× the speed you centered it. It is very annoying and occurs primarily in alt/azimuth alignment, which many of you are using; it does not occur in polar alignment.

2. Creep-after-Beep: This has been around a much longer time than any of us care to remember, resulting in slight-to-great sudden random slews in a seemingly totally arbitrary direction. You never know when this is going to pop up.

3. Sidereal Blues: This results in the ETX not tracking at full sidereal rate, whether in alt/azimuth or polar configuration; the slow drift out of the field of view can be part mechanical and a great part electronic and is easily remedied as discussed in both Parts 1 and 2.

4. Little or no Response with Centering/Tracking Speeds Less than 5 on the Autostar: I have corresponded with users worldwide who have experienced way too long periods of time attempting to fine-tune a centering maneuver with the fine adjustment Autostar speeds. Previously with my ETX-125 with a setting of 5, it took about five seconds to respond to a command to move in a direction opposite that in which the scope had previously moved; it took 12 seconds, sometimes longer with 4; and never responded with a setting of 2 or 1. Even after the motors finally responded, the reaction was a quick jerk, a motion so fast that it would incorrectly place the object on the other side of where you wanted it! Now, I get immediate, positive, and very critical adjustment at any slew/track speed.

5. Residual Drift: This occurred when you centered an object at any speed and released the arrow key(s) only to have the object continue to move (you could hear the motors continue) until you had to center again.

6. Poor Tracking: In addition to the slower-than-required sidereal speed of 3, many times tracking was very erratic in both alt/az and polar; you should expect a little more jumpy tracking in alt/az because of the nature of the moves necessary to track celestial objects.

7. Very Poor GOTO Track Record: We were all reporting misses of "several degrees" to "nowhere close."

8. Home Position Alignment Difficulty: This crucial process is the guts of how your telescope will perform in GOTO and tracking for the rest of the night; yet accessing alignment stars has been very difficult, resulting in equally poor GOTO performance.

9. And last, but not least in anyone's ETX opinion – considerable mechanical slop and inefficiencies: These issues are addressed entirely in Part 1; and the considerations/checks/adjustments that are detailed must be done prior to expecting any miracles from Parts 2 and 3. Your ETX/Autostar will not work properly if you do not read the problem descriptions and assure that you, your dealer, or Meade have made the proper adjustments to assure you that you are getting the scope and the capability you paid for.

Remember these nine issues throughout this enhancement guide.

My ETX-125 arrived with the finest optics I have ever seen for a scope of its size, but the overall package was miserably plagued with mechanical flaws. It would not clamp at all in declination (DEC) (altitude) and barely in right ascension (RA) (azimuth). The DEC axis was loose and full of backlash (over three degrees when clamped), and the scope would not keep up with objects in either alt/az or polar modes. The GOTO was miserable.

As a quick hint of how I "buffed up" my ETX and Autostar, the key is threefold and not complicated at all:

1. The mechanical aspects had to be fixed and tightened up first.

2. There is a very real and common problem after you download new software or any uploads into your Autostar. Yet it is so simple to fix that you'll hit your head in disgust. This is why so many people complain about how "wonderful my 2.0 versions worked" and why after they have installed version 2.1Ek "nothing is going right!" It turns out that it is not the download, it is not the software, and it is not the version (unless you have been living in a cave and still have the problematic Autostar version A2.3 installed). It is merely a three-step process that must be followed in sequence after your upload. Meade

don't (or didn't in the past) tell you this. The Autostar doesn't tell you this. Your PC edit screen doesn't, either. These are logical steps that we should have followed anyway.

3. Now new on the Autostar menu is an adjustment called ALTITUDE PERCENT followed by AZIMUTH PERCENT, both found by scrolling down through the "Telescope" section under SETUP. These adjustments were not there before (if your Autostar does not have them, upgrade to the latest version). They will take up your backlash, improve the response time of your arrow keys, and greatly improve your celestial tracking. These are factory-preset and should only be changed when absolutely necessary. More about that in Parts 2 and 3.

Using the steps I am about to explain, here are the results for my ETX-125 from the night of February 18, 2001, the night of a series of extreme tests after I followed the steps in Parts 1, 2, and 3. Note also that I always use polar mode and further testing is required in alt/azimuth to see if the procedures rectify the problems in that mode.

1. Out of 37 initial objects selected, including a wide range and random list of stars, planets, deep sky objects, asteroids, and comets, 31 of them were locked within the narrow (44' arc) field of view of my 26 mm eyepiece on the ETX-125. All objects I selected were at least 90 degrees from each previous object to assure an acid test over great expanses of sky.

2. Of the 31 that were in the field of view (FOV), 12 of these were absolutely dead center.

3. Six objects were slightly missed by the GOTO, all near the meridian, slewing from points far north. Even though outside the ETX's FOV, they were only out of center by about 25 arcseconds, a little more than halfway from the center to the edge of the eyepiece, and always a bit north of center.

4. To check my response time with the arrow keys on the Autostar keypad and centering capability at fine speeds, I selected 2, which I had never been able to use before, as there had never been any response before. Immediately, both motors kicked in with no backlash and no hesitation. Once the button was released, the movement ceased precisely, with no overrun as before. Immediate response/no hesitation/no backlash even at the slowest centering speed – something we've all been dreaming about.

5. Tracking in polar: At the conclusion of my tests, I wanted to check for fluctuations and oscillations in the sidereal tracking at high power (317×). I chose Saturn (at the meridian), another star on the meridian after that, and Rigel, far to the south, for a declination variance test. Rigel tracked dead center (with minor "settling-in" oscillations of only some 7 arcseconds max at first) for one hour and 7 minutes, at 317×. Saturn, higher north, and the star Castor each tracked for over 30 minutes dead center. Saturn was still going dead-to-nothing at 2:15 a.m. when clouds blocked my view for good.

6. For the final acid test, I initialized a GOTO from the star Mizar (high in the north in the constellation of Ursa Major) to Messier 42 in Orion, far south. The scope moved very deliberately with minimal noise or scatter in motion to M42, where it locked into M42's Trapezium stars at 73× dead center – again, things we've been dreaming about.

Here's how to start getting your ETX buffed and ready to do what the engineers at Meade knew all along it could do.

Your ETX-EC Mechanical and Physical Checklist

The following may seem redundant if you have had your telescope for a long period of time, but listen up. If your scope and/or Autostar has experienced any of the nine problems listed above and your frustrations are mounting, it is time to pretend the scope is still in the box and you are checking it out carefully for the very first time. If your scope is brand new, read all of this before expecting miracles out of this little blue beauty. I guarantee that if you follow the instructions and procedures through the next several pages, the reward will be getting exactly the type of telescope you thought you were getting. Spend your time observing, not fussing.

Without describing the entire construction of the telescope, I will address only problem areas that either I have located/experienced or those that have been reported to me and I have found a solution for. Note that this does not consider any aspects of the Optical Tube Assembly (OTA) other than its alignment for

Home Position (Part 3). Follow this checklist in order for easiest identification of potential mechanical problems and their easiest remedies.

Important note: Some, or all, of the following procedures may result in questions regarding Meade's warranty coverage. If you cannot do it yourself, or if you want to maintain your factory warranty for its duration, consult with Meade before making any internal mechanical (or optical) modifications, changes, or adjustments.

Declination/Altitude Drive and Clamps

Perhaps the biggest mechanical sore spot with the ETX telescopes has centered on the fork arms, the DEC/ altitude drive system, and the associated clamps for that axis. Here is the checklist for that axis:

1. The declination setting circle (on the nondriven side fork): This must be accurately set for good Home Position setup and subsequent alignment. For Home Position in alt/azimuth mode, assure that the telescope tube is level (use a bubble level on the OTA, not on the tripod). When it is level, set the circle to read 0 degrees; tighten the cinch knob as tightly as necessary to hold the circle reading in place. Many times the setting circle will move from 0 when clamping, so re-adjustment is necessary. For polar position Home Position, the circle will read 90 degrees; but I still recommend setting the circle when the OTA is level; the DEC should read 0. To hold the circle in place, put two small pieces of duct tape inside the clamp knob holding the circle so that the shiny (nonsticky) side faces the back surface of the setting circle. This allows for adjustment but also holds the circle firm over long periods of time.

2. Declination "slop" in the trunions: At the top of each of the fork arms are two beveled "trunions" that engage into holes atop the arms; there are no bearings in these holes, and there is usually a lot of slop here that will affect alignment and GOTO. With the DEC clamp engaged gently, rock the OTA back and forth and watch the tube in relation to the top of each fork; most of this play is a result of this looseness. This can be easily fixed with several (I use two) wraps of plain Teflon white tape tightly around

each of the beveled trunions as a bushing to remove this slack. Apply lithium grease liberally around the tape and then work the trunion inside the fork arm once replaced to seat and loosen the axis for use.

Another quick fix at the trunion level is to place a fiber (never rubber or metal) washer just behind the declination circle. Use one that will fit exactly inside the fork hole against the end (with the threaded hole to accept the DEC clamp knobs) of each trunion. This allows the torque of tightening to be on the washer rather than the thin setting circle itself.

3. Declination "slop" in the OTA support arms: Each side of the OTA has a support arm that connects the tube to the correct mounting fork arm. There is a considerable amount of play in all ETX scopes in these support arms. The arms fasten to the tube in the back tightly, but the front ends of the support arms are not attached in any way. The mass of the telescope allows the OTA to rock against this front surface. The OTA can be taken easily off the forks by removing two Allen screws in the rear cell of the OTA and carefully pulling straight back on the OTA. Two small clips hold the OTA support arms to the forks, so gently move the tube as it is guided out the back of the fork arm assembly. Once off, merely take some inner tube rubber repair material and glue (using rubber cement) a shape to conform to the curved front edge inside the OTA support arms. This acts as a tight but soft bushing to give strength and support to the OTA when it is back in place and reduces up to one degree of DEC/altitude slop from your system. Be sure, at this point, to hold a bright light behind each of the four OTA mounting holes in the OTA support arms and look for tiny hairline cracks, which are very common; if the plastic OTA support(s) is cracked, contact Meade immediately for a replacement. Do not over-tighten the bolts in these holes when replacing the OTA.

4. Declination clamps: Particularly troublesome with the larger ETX-125 is the failure to firmly clamp both the declination (alt) and RA (azimuth) axes to where they (1) will hold the telescope in place without slipping and (2) allow the motors to properly slew the telescope without the clutch slipping.

If your clutch on either axis slips during slewing, your GOTO will be much less than perfect. The telescope encoders "think" the scope is moving to the proper location because the motors are turning the gear works; but if the scope is not moving with the gears, it will be out of position. The same is true for tracking; if either clutch (both for alt/az mounting or just the RA for polar) is slipping badly, your telescope will slip when the weight of the telescope assembly is heavier opposite the direction of sidereal motion. Thus, the telescope will not keep up with the objects as the Earth turns.

This is a problem that must be fixed prior to going any further. If your clutches are slipping, the telescope electronic functions cannot work. It is as simple as that. You can fix them; usually all it takes is being brave enough to get in there and simply degreasing all the surfaces that should be dry and clean to grab the gears for slewing. You will find that every ETX has way too much lubricant on both axes. If the lubricant has seeped between the drive gear and the clutch plates (which it will), then your GOTO may be error-prone.

If you find that you are having to really bear down on either or both clamps to get the scope moving from a dead stop immediately, you are clamping too hard and might damage the telescope mounting. It is time to clean the clutches. (Electronic fixes discussed in Part 2 will help in this regard as well.)

Although it may seem intimidating at first to get into the DEC and RA motor assemblies, it actually is not; and the benefits are tremendous. Follow this procedure step-by-step to allow full use and precision of the declination clutch and both DEC and RA drive systems. Degreasing is the key here. If you degrease with mineral spirits and then carefully and sparingly use lithium white grease only on the gear teeth (not on the flat clutch/gear contact surfaces), your tracking, clamping, and GOTO are immediately improved and you are ready to move on.

Right Ascension/Azimuth Clutch Clamp and Drive System

Be sure to use care throughout your work and exploration here, and you will be able to adequately adjust the drive mechanisms to reduce play and mechanical backlash. Autostar/electronic backlash is also a factor, and its fix is discussed in Part 2. You must do the mechanical fixes first.

Perhaps the most important factor regarding tracking

and GOTO accuracy from a mechanical standpoint is that the clutch of the RA drive is prone to easy slippage because of the large torque of the system that rests primarily on the pivot axle. There are wires, those that run to the declination drive and the ones for the battery housing connected to the rear plate, that you must remove to access the RA compartment. These are extremely fragile and must be handled with the utmost care.

To remove the clutch plate requires that you (1) remove the chrome RA clamp from the top turntable using the very small Allen screw (exercise caution so as not to strip the head of the screw); (2) use the clamp knob as a tool to loosen and subsequently unscrew the long bolt that runs down into the compartment and ultimately engages with the clutch plate; (3) once loose enough, finish detaching the clutch plate by hand until it is loose enough for resurfacing the side that presses against the drive gear.

You may leave the clutch plate exactly in position to service it. I have found that using coarse sandpaper, followed by medium grade steel wool eliminates the dimples and prepares a uniform, yet gripping surface. Protection of the internal electronics and small gears is essential.

Once the clutch plate is disengaged, the large metal flat drive gear can also be pulled away from the axle and cleaned thoroughly.

The important step here is cleaning, just as with the declination fix. Clean the following thoroughly with mineral spirits after preparing the surface of the clutch plate: (1) the clutch plate, both surfaces; (2) both sides (flat) of the metal drive gear; (3) the surface of the axis against which the drive gear locks (opposite the clutch plate from the gear).

Once cleaned, use a toothpick to spread some lithium grease (white) along only the teeth of the large gear, making sure to not get the grease on the smooth flat surfaces. Re-insert the gear onto the axle snug against the back (top from the scope) surface plate; this will engage the gear tightly within the worm gear. Then simply put the clutch plate back, with no lubricant at all, so that the grooved fitting engages properly for alignment. Tighten finger-tight. At that point, go back to your clamp tool and tighten reasonably tightly until it is all firmly snugged together. You finish tightening after you have the base plate back on and are ready to upright the telescope. You also may adjust your clamp at this time.

You now have a very tight RA clamp that allows firm clamping without over-tightening and one that will allow long and accurate slews without slipping.

Remember: if your clutches are slipping, your GOTO and Autostar will not work up to your expectations. Indeed, if there is grease in the clutch systems, the problems will merely get worse, forcing you to clamp even harder as the grease spreads, resulting in more and more inaccurate tracking and GOTO functions. So do these precautionary steps before doing anything else.

Before any testing and before any substantial movements using the gears in combination with the Autostar handcontroller, you must train the motors in the mode (i.e., polar or alt/az) that you will be using primarily. Failure to train at this point will result in inaccurate GOTOs and sidereal tracking rates.

Once you have aligned, cleaned, degreased, and loaded your DEC and RA drive systems and once you have properly trained with the Autostar, you can test it out indoors. Merely clamp very lightly, do a dummy initialization, and try the arrow keys to slew. The less you have to tighten either clamp, the better job you have done. After setting and cleaning the clutches, your clamping pressure likely will be far less than previously required. Only a true GOTO will tell how well you have eliminated mechanical slop and backlash.

You can also check the RA tracking accuracy in polar mode or the combination alt/az tracking accuracy in the alt/azimuth mode by initializing and doing a fake indoor alignment after you have properly trained the motors. Once alignment is successful, the ETX motors should engage into the tracking mode. Monitoring the motion of the telescope relative to its setting circles will tell you the consistency over a period of time of tracking. Check it at 10 or 20 minute intervals and note the motion. Then begin anew for another identical period of time and note the lapsed motion relative to the setting circle(s).

Now that we have your telescope mechanically ready for the precision it can provide, let's examine your Autostar, updates on software versions, and all subsequent initializations and preparations that are necessary to take your telescope outdoors for its perfect night.

Part 2 – Autostar Downloads and Post-Download Initialization

Thus far in this performance enhancement guide, we have discussed the critical need of having an optimum mechanical operating system for the computerized ETX telescope to enable it to function properly with the Autostar GOTO. Without a responsive, tight, and nonslipping telescope, all the computer software and programming knowledge in the cosmos will not help you slew from one object to the next.

Before proceeding to optimize your telescope through these "performance enhancement reports", make sure that your telescope has been thoroughly checked for all mechanical pitfalls that can be detrimental to the scope's performance and to your enjoyment of this hobby.

As mentioned previously, our problems with the ETX are a complicated combination of mechanical problems and glitches that can be fixed (Part 1); improper initialization procedures in downloading and post-downloading (discussed here); and user error, including alignment, calibration, training, and leveling the telescope for perfect performance.

Downloading and Uploading Software

I will be the last to pretend that my downloads of new Autostar data and versions from Meade have been fun and easy; had it not been for the help of fellow-ETX-ers Dick Seymour and Mike Weasner and others on the ETX Web Site, I still would not understand the errors of my ways.

The purpose of this series is to instruct you on ways to overcome the problems that we have all experienced in the past and tweak your telescope to perform as it was designed. If you don't believe that such miracles happen, merely go back and read my test report on the night when all of my modifications, adjustments, and inputs were finally corrected and/or added to the telescope/Autostar combination.

I will leave downloads and data editing to the computer experts among us, as I certainly am not among them. What I want to convey is the process after you have edited the libraries or new ephemerides into the latest version of Autostar that you have uploaded from the Meade Web Site.

It is in this process following the Autostar upload that we have been introducing what appears to be conflicting information to the Autostar, which may be a culprit in such things as the "rubber band," improper tracking, random slews, and much difficulty in GOTO and alignment. Time will tell but by following a simple procedure and adding one more sequence to your post-download commands to Autostar, I believe that your ETX will suddenly become the "buffed" telescope it is supposed to be.

It is important that after upgrading to any new version, whether it is a new Uploader program (such as A2.4) or software/database/ephemeride (such as v2.2Er or later) or just a new ephemeride for a newly discovered comet, you must reset your Autostar after the first initialization is finished (this would be the initialization that automatically commences after "Download Complete/Initializing Autostar").

The following is the sequence that I have discovered that allows considerably better communications between the Autostar and the ETX. I do not pretend to know why or how it works; I merely know that it does work, as will be seen in my carefully controlled testing parameters. But first, let's review the sequence as I believe it should progress:

Post-Uploading Sequence for Autostar and Telescope

This by far is one of the most important, combined with proper mechanical behavior, of getting your ETX and Autostar to perform properly. We are set in our ways for sure and usually unless we are told to do something through instructional information, we typically will not do a logical step or often, as in this case, not think it important.

I stumbled across what may be the major factor in many computer flaws that users are reporting quite by accident: when my initialization sequence began after "downloading complete", I never heard the motors being calibrated. That, when the Autostar is RESET, as it should be after every download, is one of the first processes that should take place with initialization.

This section addresses two new avenues that are open to you to improve your telescope's computer performance: (1) the post-software upload initialization process that you must follow after upgrading your Autostar and (2) adjustments of the "Altitude Percent" and "Azimuth Percent" values prior to motor training that are available on the Autostar to remove backlash as well as improve response time when using the slow and fine rates of your arrow keys. This, as you will see, is a major factor toward getting your Autostar to GOTO as well as track more efficiently.

Post-Upgrade Readying Sequence

It is exciting to download new data from the Meade Web pages that you expect will improve the accuracy and enjoyment of your telescope; at the present time, Meade have done an excellent job of providing faster and simpler methods of data-sharing with the ETX users than ever before. However, it is not the purpose of this document to discuss downloading from the Meade Site nor uploading into your Autostar. What will be discussed are the many steps after that familiar "Download Complete" message appears on your PC (or Macintosh) once the Autostar has sucked all the information from your files.

Notice that immediately after "Download Complete", your Autostar shows the initialization sequence, usually repeating (1) the Sun warning until you tell it to go away. From there, you go to (2) the instruction page, next to (3) the date, (4) local time, and on to (5) "Daylight Savings?" and off to (6) "Afghanistan" probably, then to towns in your state or province, and finally to (7) to "Telescope Type."

Note: on versions other than 2.1Ek, the initialization process includes as the first step (before Sun warning) your selection for language; at the present time, there is no prompt for this as the latest versions have been English only. Once your selection of language is made, "testing motors" appears; and there is slight movement (calibration) in both altitude and azimuth axes.

Once all those parameters have been entered, you turn off the telescope power, disconnect the cables from the Autostar and telescope, turn off power to your computer, disconnect cables from its port, and you are ready to go outside and align, right? Wrong.

Figure 6.13 shows the post-download sequence that you should follow to assure optimum control of the ETX.

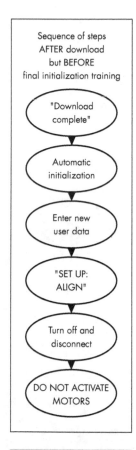

Sequence of steps AFTER download but BEFORE final initialization training

"Download complete"

Automatic initialization

Enter new user data

"SET UP: ALIGN"

Turn off and disconnect

DO NOT ACTIVATE MOTORS

Figure 6.13.
Post-Autostar download.
Courtesy P. Clay Sherrod.

Pre-Motor Training Initialization Sequence

Now you have (hopefully) downloaded your new information into the Autostar, disconnected everything, and are ready to go out for an evening's observing to "test the new stuff." In the past, such evenings have been major disappointments to many observers, with new idiosyncrasies popping up all over the place: "rubber banding," random slewing, poor tracking, horrible alignment.

Remember that your Autostar has already initialized itself immediately following its last download but it did not RESET itself. That is something you must do to remove the previous data on motor training. It appears that your last motor training is not erased like other data (your name, site location, telescope type, etc.); indeed, the difficulty of locating objects with the Autostar suggests that a data recognition problem exists in its memory, likely caused by two conflicting drive motor training sequences, with which the Autostar must reckon the best it can.

So here are the additional steps that you should take prior to training your motors outdoors and of course prior to aligning for the night's use. Some of these appear repetitive but are necessary because you are clearing out your entered data from the Autostar to give it a fresh start after installing the new versions of the software. Forget the fact that an initialization has already taken place; you are about to do another one (following along using Figure 6.14).

Once you have finished this last series of preparatory steps, you are getting close, but are not there yet, to actually training the motors (see Part 3) and aligning your telescope for operation.

There is one more very important and helpful process that you must do prior to training the motor drives of your ETX.

Setting your Altitude and Azimuth Backlash Percentages

New on the latest Autostar uploading tool (A2.4 and beyond with 2.1Ek and above software) is a wonderful addition that eliminates sloppy tracking, problems with reaction time using the fine tracking arrows, creeping

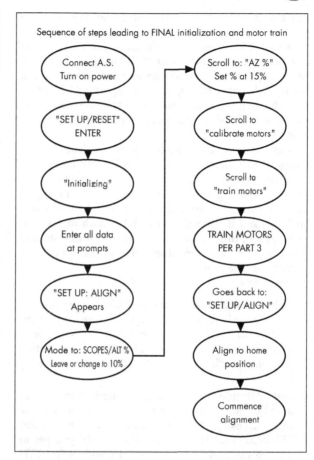

Sequence of steps leading to FINAL initialization and motor train

Connect A.S.
Turn on power

Scroll to: "AZ %"
Set % at 15%

"SET UP/RESET"
ENTER

Scroll to
"calibrate motors"

"Initializing"

Scroll to
"train motors"

Enter all data
at prompts

TRAIN MOTORS
PER PART 3

"SET UP: ALIGN"
Appears

Goes back to:
"SET UP/ALIGN"

Mode to: SCOPES/ALT %
Leave or change to 10%

Align to home
position

Commence
alignment

Figure 6.14. Post-Autostar download – initialization steps. Courtesy P. Clay Sherrod.

after slewing, sidereal oscillations, and general gear train backlash.

Backlash is simply the residual time necessary for the gear train (the gear teeth in particular) to catch up and reverse any motion. If you have been going left and suddenly wish to reverse and go right, it takes some time for the gears to catch up with each other to the point where torque is finally once again applied to the actual drive gear (as discussed in Part 1). When using the very slow arrow keys with Autostar, sometimes there is no motion whatsoever because of the great time necessary to turn the combination of so many gears. Hence, a lot of users have told me that they never even used 1 or 2 because nothing happened!

Believe me, if you follow all of these enhancement techniques, you will be able to effectively use speeds 1 and 2 from now on, with little or no hesitation before and no residual creep after you press the button.

Every Meade telescope is different and each has its own distinctive combinations of RA and DEC backlash. For example, my RA backlash was horrible when I first got the telescope; taking the base apart and cleaning and adjusting the clutch and drive system (see Part 1) greatly helped reduce the backlash but did not remove it altogether.

Conversely, my DEC axis had no backlash at all, even though it had mechanical "play" in its components.

So, mechanically, I have my telescope as good as it can be without going into custom drives and super-charged custom overhauls. But it still had tracking problems and it still suffered with backlash in RA. Just in the nick of time, a warrior arrived on a white stallion with the Autostar improvement: the Alt and Az percent adjustment function.

Here's how to get yours started.

1. Make an initial setting prior to Train Motors (Part 3) on either or both axis percentages. If one (or neither; you would be blessed if that was the case) axis appears not to have backlash problems and drifting problems, just leave the two functions alone. We envy you.

2. Both Altitude Percent and Azimuth Percent are found on the Autostar keypad (with A2.4 and above with v2.1Ek and later software) under SETUP/ TELESCOPE/MOUNT. Scroll down until you get to the two categories. Do not confuse with AZIMUTH RATIO and ALTITUDE RATIO, as these are factory presets for the engineering specs of the motor system and should never be changed. In my opinion, the ability to change these options should not even be accessible to the user. Altitude Percent and Azimuth Percent are found after the ratio categories.

3. When either appears on the lower line of the Autostar readout, press ENTER to bring to the top and open; both are factory set at 01%, allowing for no restriction on backlash whatsoever. Meade suggest that if you are having a backlash slop problem, start with 50% and work your way back down to 0 percent. I don't recommend this for two very clear reasons:

   (a) At 50% the motion of pressing the arrow key(s) would be very jerky, with the telescope drive suddenly surging ahead and then slowing to the normal rate (1–9) you have selected. The 50% actually is a speed surge of 50% faster for the first 0.3 seconds (roughly) than the rate selected to compensate for delays of backlash to allow the gears to engage much more quickly. Once the backlash is removed by this surge, the rate of slewing/tracking is reduced to the normal rate selected.

(b) You should have the backlash setting at a reasonable percentage for correcting your scope prior to Train Motors, so that training will take the backlash percentage into account when "marrying" the Autostar to the telescope drives.

Thus, I offer this suggestion that has worked perfectly for me. If you are having backlash and/or lagging problems in both DEC and RA, even if it seems severe, try a setting of 15% for each axis (each must be set separately on the Autostar). This should do the trick; I have found that even 20% is too high and unpredictable, and above that is not usable. If, on the other hand, your primary lag and response time/ backlash troubles are in the RA/azimuth axis only, don't even bother with the DEC/altitude setting. Merely set the Azimuth Percentage function to 15%. I think that value will cover nearly all ETX telescopes.

Once your percentage has been keyed in, press ENTER and it remains until you once again RESET the Autostar. Now you are ready to Train the motors and align the scope precisely for the "Perfect GOTO Experience." Remember, be sure to always Train your motors with the telescope in the configuration (i.e., polar or alt/az) that it is going to be used in and always set that configuration (under SETUP/TELESCOPE/ MOUNT) to whichever you are using before training the drives. Failure to do so is another pitfall that some users have fallen prey to that results in less-than-perfect performance

Appendix: The Test Parameters for Autostar

Figure 6.15 shows the three parameters in evaluating my conclusions regarding Autostar accuracy: the need for a RESET function after downloading, the resulting performance increase by changing the backlash para- meters, and finally, the decision that changing the

Clay Sherrod's AutoStar Test
Arkansas Sky Observatory Week of Feb. 9-16, 2001

AUTOSTAR #1	AUTOSTAR #2	AUTOSTAR #3
PRIMARY AUTOSTAR	BACKUP AUTOSTAR	#2 NOW CHANGED
Version: v2.1ek	Version: v2.1ek	Version: v2.1ek
Loaded: Feb. 14, 2001	Loaded: Feb. 13, 2001	Loaded: Feb. 13, 2001
using A2.4	using A2.4	using A2.4
DATABASE:	DATABASE:	DATABASE:
158 asteroids	158 asteroids	158 asteroids
201 satellites	201 satellites	201 satellites
144 comets	144 comets	144 comets
5 tours	5 tours	5 tours
0 user objects	0 user objects	0 user objects
VARIABLES:	VARIABLE:	VARIABLE:
Autostar Reset	*Autostar NOT Reset*	*Autostar Reset*
after download &	after download &	after download &
1st Initialization,	1st Initialization,	1st Initialization,
but BEFORE alignment.	but BEFORE alignment.	but BEFORE alignment.
Azimuth Percent	*Azimuth Percent*	*Azimuth Percent*
changed to 15%	left at factory	changed to match
Altitude remains @1%	setting (1%)	AUTOSTAR #1
Training:	*Training:*	*Training:*
Trained after download	Training same as #1	Exact same as
and AFTER RESET	but after download,	#1
	NO reset	

backlash percentages should be made prior to Train Motors.

I use two Autostars, one for actual operation at the scope and the other to download new data and Autostar versions. Also, the second unit has been invaluable at helping unravel what must be done electronically to make the computerized Meade telescope perform electronically the way it was designed.

Figure 6.15.
Autostar test results.
Courtesy P. Clay Sherrod.

Autostar #1 is the primary Autostar and you can see that it has been loaded with everything pertinent that has been discussed so far. It is the one that provides the incredible results demonstrated in Part 1. Autostar #2 was similarly downloaded as can be seen but had not been RESET nor had the azimuth/altitude percentages been updated. AutoStar #3 is merely Autostar #2 after resetting it and programming in all the parameters that made AutoStar #1 function as well as it does. The complete listing of variables and parameters is below each one.

My conclusions are based on these Autostars and the variables, from the following results:

1. All three used the A2.4 updater and v2.1Ek software loaded on the same date from the same computer. With the parameters shown for Autostar #1, tracking was near-perfect; the GOTO (described in detail in Part 1) speaks for itself, and all delays seem to be eliminated; response is immediate and there is no residual overcorrection at any slew/center speed. In essence, this combination on my ETX-125 gave near-perfect performance in every respect, far better than I had ever experienced with the ETX. Setting the Azimuth Percent (no need to change the Altitude for my telescope) took out virtually every bit of backlash that my mechanical adjustments could not. Alignment was quick and accurate, even in polar mode.

2. Everything in Autostar #2 is identical, except the unit was not Reset after the first initialization immediately following downloading of v2.1Ek. Nor were Azimuth or Altitude Percent changed. The results were identical to what everyone is experiencing: creep-after-beep, wild slews to alignment stars, imperfect GOTOs, sidereal tracking problems, and significant rubber-banding in alt/azimuth mode. Perhaps the biggest problem was the ineffective slow and fine settings for centering (numbers 1 and 2 on the keypad) with bothersome waits to slew and then residual travel after the object was centered; this was eliminated with the settings in Autostar #1.
    This comparison was done on the same telescope. Only the Autostar information was different; both Autostar handboxes were trained with the ETX-125 as well, so only the variables could be a source of change.

3. Then, I took Autostar #2 and turned it into Autostar #3 by redoing the reset, then a new initialization,

added the percent parameters, and subsequently trained the motors. The results? AutoStar #3 performed 1-on-1 with Autostar #1 with the identical and newly loaded parameters that Autostar #1 had had all along. This merely proves that it is: (a) not the telescope, since the same ETX-125 was used for all tests; and (b) not the Autostar, since Autostar #2 failed miserably, but when reprogrammed to the configuration of Autostar #1, performed brilliantly!

We have seen the advantages of providing the best mechanical performance possible for our ETX scopes and also the importance of the crucial sequence necessary for a clean upload after new software is installed. We have seen that a RESET might be the easiest way of fixing all your errors even though none of the literature or manuals (through version 2.1Ek) suggested it should be done.

Part 3 – Training the Drives and Celestial Alignment

For the most part, the "performance enhancement" for your ETX is now complete as far as the telescope goes. You have remedied mechanical shortcomings and you have followed a sequence for download/reset/initialization that leads you to training the drives. From here on out, the entire success of your sky searching and tracking proficiency is entirely up to you. You are in control; any problems at this point are chalked up to "user error."

For this final section, I want to walk you through the process of a very precise Train Drive procedure, followed by determining the Home Position in both alt/azimuth and polar configurations. Although there is much said about this on the ETX Web Site and some in the Meade instructions, there are still some guidelines and helpful hints to let you do it better. We are, after all, after perfection.

Training the ETX Drives

Important: You should always train the drives in the configuration in which you plan on using your scope the most, whether it be alt/azimuth or polar. You must

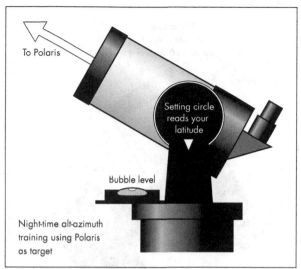

Figure 6.16.
Alt-azimuth training.
Courtesy P. Clay Sherrod.

first select TELESCOPE/MOUNT and either alt/az or polar. Press ENTER and then scroll to Train Motors. This increases the ultimate slewing/tracking precision of both drive trains.

1. Before starting to train the drives, make sure that your clamps are fully engaged for accurate movement of the motors and telescope during the training procedure. Failure of the clamps to hold tightly will result in a lack of anticipated movement by the Autostar and hence failure to obtain an accurate motor training.

2. Select your target object, for example, the North Star Polaris for alt/azimuth (see Figure 6.16) for training at night or a distant elevated terrestrial object for day-time polar training (Figure 6.17). With the lowest magnification, center the target in the main telescope eyepiece; gradually increase the magnification until the highest possible is achieved and continue to center the object dead center each time. A reticle eyepiece greatly assists in the centering process for Training but is not necessary.

3. Remember that in training, the Autostar will move the telescope first in two opposite altitude directions and indicate an arrow key to push to recenter the object after each move. Make sure that you use the lowest practical slewing speed (2 if possible) to recenter the object once moved. This is because you

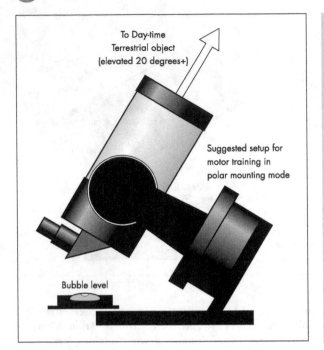

To Day-time
Terrestrial object
(elevated 20 degrees+)

Suggested setup for
motor training in
polar mounting mode

Bubble level

Figure 6.17. Polar training. Courtesy P. Clay Sherrod.

only have one chance to center. If you overshoot the target (too much past center), you cannot reverse the direction in the Training function.

4. A common mistake made by many users that I have talked with is Training azimuth twice. This is easy to do and many times results in some of the Autostar oddities that can occur. Here's how it works and how to remember to not do azimuth twice in a row:

(a) When you scroll to TELESCOPE/TRAIN MOTOR and press ENTER, instructions commence about the training process. Press ENTER.

(b) The first to appear will be AZ. TRAIN. Press ENTER, at which time the telescope will move slightly in one azimuth direction and prompt you to CENTER; do so and press ENTER.

(c) The telescope will slew in azimuth in the opposite direction and again stop and ask you to CENTER and press ENTER; do so.

(d) When you press ENTER after the second azimuth slew centering, rather than defaulting to the next procedure (which would be to Train in altitude), here is what shows up on the screen: AZ. TRAIN, just as before. If you hurriedly hit ENTER, you will be retraining in azimuth, not starting what you should be doing, which is training in altitude.

(e) So, instead, when the AZ. TRAIN prompt comes up after you have already done so, use the SCROLL key to move down one step to ALT. TRAIN and press ENTER. Again it will begin slewing in one direction, this time in altitude and ask you to recenter the object and then repeat in the opposite direction.

You would be surprised how many ETX users out there have a double azimuth training because of this simple-to-correct error.

5. Important: Before commencing training, be sure to get to SETUP/TELESCOPE, select TELESCOPE, scroll down to MOUNT, and select it. Then, under MOUNT scroll down to ALT/AZ or POLAR and select whichever one you plan to use by pressing ENTER when that function is on the top line of the Autostar screen.

You are now ready to Train Motors for your respective setup (i.e., either polar or alt/az). As previously noted, this is a very, very important process that will assist in greatly enhanced performance of your telescope's computer capabilities. If you do not train properly and very deliberately and precisely, you cannot expect the ETX and Autostar to do what Meade claim it will do (and it will do it if everything is set up properly).

Training in Alt/Azimuth Mode

Training in alt/azimuth mode can actually be more accurate than training in polar mode, even though in my opinion the overall tracking/slewing performance in Polar is ultimately superior. However, remember that if you plan to use polar mode, you must Train in polar mode.

In the alt/azimuth configuration, training is made more accurate and much easier by using the bright North Star Polaris as your training object for several reasons. (Note: never use Polaris for polar mode training. Use a distant and elevated terrestrial object.)

1. Polaris, for the duration of the training exercise, is relatively fixed in the sky and therefore does not

appear to move, making it an ideal target at extremely high magnification.

2. It is a point source of light and very distinct, thereby making it easier to place it "dead center" in your eyepiece.

3. Since some elevation of the OTA is required during training to provide a slight degree of gear loading, Polaris is elevated from the north horizon to your latitude, probably at a higher angle than most terrestrial objects at any distance.

4. Terrestrial objects are subject to washout at high magnifications. At low angles this presents a fuzzy and vague image of the tiny detail you wish to center each time (e.g., an insulator atop a distant electrical pole).

Figure 6.16 demonstrates clearly the position of your ETX telescope and base for training purposes. Note that this is not the Home Position. Be sure to take your time. Do the training using the overall steps above and you will have much success in matching the Autostar recognition to your drive train. The results will be worth the time and effort.

Training in Polar Mode

Polar mode training should always be done in day-time for convenience and ease. Never use Polaris as you would in alt/az training, since the azimuth train procedure would result in little perceptible offset when moving the telescope for the procedure and thus result in very poor motor training.

Figure 6.17 demonstrates the proper orientation (ideal position shown) for polar mode training. You may not be able to reach the high altitude demonstrated in this drawing, so certainly lower the OTA to access the best object.

Other than the time of day (or night) and the telescope orientation, there is little difference between polar and alt/azimuth motor training. Just remember to set the Autostar to POLAR mount if you are going to train in polar mode and use polar mode before proceeding to Train Motors.

Important note: It appears that in some versions of the Autostar if you change the date during the Autostar initialization after the very first setup, the selection of polar mount configuration will not be defaulted to on subsequent dates. In other words, if you kept entering "May 12" as your date having already reset the Autostar to POLAR, then "polar" would be maintained as your mount configuration of choice. However, once you key in tomorrow's (or any subsequent) date, the Autostar may automatically default back to "Alt/Az". This is a common and critical error of many users of not checking the mounting mode before alignment each night. I recommend getting into the habit of checking what mount configuration is selected in the Autostar before any alignment procedure even begins.

Maintaining and Retraining

If you are not satisfied with the way the training went, do it over again. Push RESET under the SETUP category and start over. Once the training is complete, you should not have to retrain (and always RESET first) except when:

1. you upload new data/software/ephemeredes onto the Autostar from your computer;
2. you have had an Autostar failure and need to reboot;
3. you have moved the Autostar to a telescope other than the one it was trained on, or you have changed Autostars to use one not matching the current telescope;
4. your slewing/tracking seems so far off that you wish to RESET and retrain to start anew.

If you are having Motor Unit Failure or just odd behavior in the motor function, always attempt a CALIBRATE function before going through the rigors of motor retraining. The motor calibration function will allow "syncing" of the motors quickly and a retrain may not be necessary.

Answers to the most commonly asked training questions are:

1. You do not need to retrain if you change locations. However, if you have several locations (up to five) keyed into the Autostar with exact latitudes and longitudes, be sure to SELECT the exact location you are observing from to improve the GOTO capability of the Autostar.
2. If you work on your drive train and/or clutches at all, even to degrease (as described in Part 1), you must

retrain the drives each time you mess with the drive system.

3. If you clone an Autostar containing your latest Motor Train to another Autostar, the newly cloned unit must be trained to the telescope it is to be used on. If the second Autostar is to be used with the same telescope, there is no need for retraining.

4. As stated above, you must retrain after every Autostar upload or every RESET.

5. If you choose to change, via the Azimuth Percent and Altitude Percent Autostar function (see Part 2), you do not need to retrain the motors.

6. If you decide you want to change from, say, polar to alt/azimuth mount configuration permanently, you should RESET and retrain the motors. Be sure to go into the Autostar menus and select the new mount choice as well.

7. Alternately, if you are using, say polar mode and want to just do a quick observing session in alt/azimuth configuration, you should not retrain the motors. However, for proper tracking you would still have to put the Autostar into the alt/az mount function.

Home Position and Celestial Alignment

This final section demonstrates some of the most common causes of mistakes and mis-slews that result in poor tracking/slewing/GOTO operations. Putting the telescope in the Home Position and then subsequently properly aligning are the two most crucial steps to accuracy and enjoyment of your ETX. Curiously, these are the two that are more prone to user error than anything discussed until now. So let's do it right the first time. You have spent all this time checking for and correcting mechanical problems that prevent accuracy; and you have properly installed, re-initialized, and trained the Autostar to sterling quality. Now let's get that telescope set up correctly and you "have ignition!"

Starting with the Home Position, this is perhaps the most critical step to success; you must "start right to end right" as they say. It is very important that you start in this position every time you take your telescope outdoors for observing. Failure to do so will result in disastrous tracking and possibly even damage to the telescope motor units. For the most part, it is far easier to attain the Home Position in the alt/azimuth mode than in polar, hence the reason that the majority of ETX users prefer that configuration. If done correctly, the telescope will function adequately in either mode, although polar is preferred for astrophotography and long-duration high-powered viewing.

Home Position for Alt/azimuth Alignment

Figure 6.18 demonstrates the correct orientation for the ETX to achieve proper Home Position.

Follow these easy steps to achieve perfect Home Position in alt/azimuth mode:

1. Level the tripod or observing surface as best as possible and place the telescope onto it.
2. Orient the telescope and mount (tripod or table) so that the control panel (where the Autostar plugs into) faces west.
3. Using a bubble level, first attempt to level the flat turntable (the top plate of the ETX base that has the azimuth clamp in it) in all directions. Some variation can be expected, but do the best you can.
4. Now unclamp the azimuth axis and rotate the telescope counterclockwise all the way to the hard stop and back clockwise until the fork arm with the declination circle is over the control panel and the optical tube assembly faces near north. Attempt to do this at dusk or night-time if possible so that you can see Polaris.

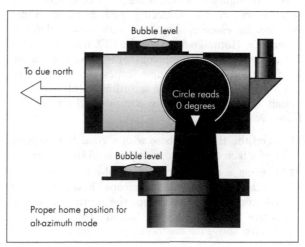

Bubble level

To due north

Circle reads 0 degrees

Bubble level

Proper home position for alt-azimuth mode

Figure 6.18. Alt-azimuth home position. Courtesy P. Clay Sherrod.

5. If Polaris is visible, unclamp the altitude axis and rotate the telescope in both axes to center the star in the telescope eyepiece. This allows you to accurately reckon very close to true north. Once Polaris is centered, lock the azimuth axis and rotate the telescope downward to near level. Your telescope is now pointing north and near the Home Position.

6. At this point, do not change the azimuth position but leave it securely clamped. Using a bubble level, level the OTA carefully as shown in Figure 6-18 and clamp the altitude clamp securely when the OTA shows perfectly level. Do not rely on the declination setting circle at this point.

7. Once level, check the declination setting circle. If it is set correctly, it will read 0 degrees. If not, gently turn it (you may have to unclamp the knob in front of it a bit) until it does read that setting.

You are now in the alt/azimuth Home Position and are ready to align your telescope to the sky.

Home Position for Polar Alignment

For very accurate tracking at high magnification and perhaps better mechanical response, many choose polar configuration for their ETX. Although more precise, the overall attaining of Home Position is also more difficult to achieve. Comparing, it takes at least three or four times longer to precisely achieve Home Position in polar than it does in alt/az mode.

Note in Figure 6.19 that there are a couple of extra steps necessary for achieving perfect polar Home Position. In essence, for those with some telescope experience, Home Position for the ETX telescope is no more than the polar alignment that we have been doing for years; but with these new scopes and the built-in safety stops, it must be done according to this telescope design. Follow these steps for accurate and quick polar Home Position:

1. Level the tripod or base with a good bubble level and attach the telescope to this level mounting.

2. If using a tilt-head tripod, tilt the top plate (that which will hold the telescope base) and the telescope fork arms toward the north, roughly to where you believe your latitude (in degrees) is located above the northern (or southern if south of

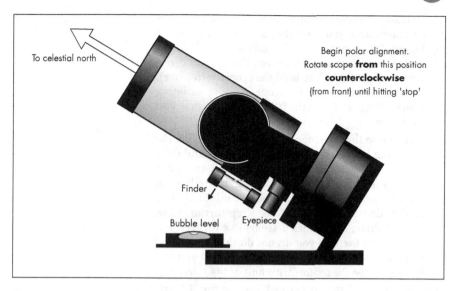

To celestial north

Begin polar alignment.
Rotate scope **from** this position
counterclockwise
(from front) until hitting 'stop'

Finder

Bubble level   Eyepiece

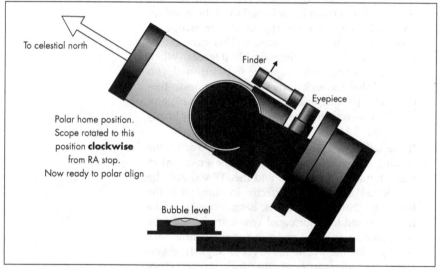

To celestial north

Finder

Eyepiece

Polar home position.
Scope rotated to this
position **clockwise**
from RA stop.
Now ready to polar align

Bubble level

Figure 6.19. Polar home position. Courtesy P. Clay Sherrod.

the equator) horizon. If northern, this position would be roughly marked by Polaris, the North Star. Many tripods and wedges have a latitude grid marker for you to roughly preset to your exact latitude prior to fine adjusting. Then aim the entire telescope/tripod assembly where the fork arms are pointing to an elevated due north position.

3. Unclamp the telescope in both the RA and DEC axes and move until you reach the odd position shown in Figure 6.19 (top), with the eyepiece and finder under the telescope (be certain the eyepiece set-screw is tight enough to hold the eyepiece in place). The lens end of the telescope is now pointing toward Polaris, or north. It is important that you start here.

4. Now clamp the telescope firmly in DEC, leaving the RA clamp loose. Gently rotate the RA axis and OTA counterclockwise until you hit a built-in "hard stop" (which prevents wire winding in the base of the telescope). Note: the ETX-60AT and ETX-70AT models do not have hard stops. Keep turning until you get there; sometimes it seems like forever, but you will get there. If you do not do this procedure, you will not be able to computer-align and you will likely damage the motor/drive unit of the telescope.

5. Once the hard stop is reached, reverse the RA axis rotation slowly (turn it clockwise) until the eyepiece and finderscope are on top of the telescope as shown in Figure 6.19 (bottom). This amount of rotation will be approximately 180 degrees. When the eyepiece and finderscope are so positioned, you will find that the fork arms are aligned one-to-the-other and parallel to the ground (Figure 6.20) and should be exactly in the position shown. Now firmly clamp the RA axis.

6. If the telescope's setting circle is set correctly, the reading in Figure 6.19 (bottom; Home Position) in DEC should be 90 degrees and the OTA should be positioned as shown, perfectly in line with the length of the two fork arms. Back away from the telescope and look for equal symmetry of the tube compared to one fork and you can determine very closely with the eye if the OTA is slightly askew from the alignment of the fork arms. Once alignment is achieved, clamp the DEC axis securely.

All of numbers 1 through 6 can be done prior to nightfall. The following must be done after it is dark enough to see the star Polaris with the naked eye.

7. Without unlocking either axis, move the telescope and tripod (use the fine adjustments if angles are minor or if your mount/tripod has them) until you can get Polaris in the finderscope. Do not unclamp the axis.

Figure 6.20. Polar home position 2. Proper positioning for the fork tines in polar mode. Note that the ends of the forks should be in a line parallel (as shown with arrow) with the level ground. Courtesy P. Clay Sherrod.

8. Using only the slow motion controls of the tripod (do not unclamp either axis), adjust the image of Polaris until it appears in the field of view of any eyepiece that provides about one degree field of view (FOV). This is very important for precise celestial alignment.

9. Once you attain a one-degree (or close) FOV, you can now offset Polaris in the telescope to obtain a precise celestial pole pointing. Polaris is not true celestial north, which is about one degree away. That one degree can make all the difference in the world in the accuracy of your GOTOs and in your overall sidereal tracking over long periods of time. Remember this rule: True celestial north is about one degree from Polaris in a line that extends nearly through the star at the opposite end of the little dipper handle. This star is the second brightest star in Ursa Minor – named Kochab. Being circumpolar, this star should always be visible even when low in winter months for most locations in the northern hemisphere.

10. Using the tripod or wedge adjustments only, you want to *raise* the telescope assembly in the eyepiece FOV until Polaris is at the very edge of a one-degree

FOV eyepiece. Now the idea is to move the telescope assembly until it is offset that exact one degree from Polaris toward Kochab. To do this merely slowly tweak the slow motion controls of the tripod or wedge until Polaris reaches the opposite edge of the FOV in a direct line toward Kochab. Once on this edge, the telescope is offset about one degree from Polaris and very close to the true celestial pole.

Initialization Checklist

Once you have achieved Home Position in either mode, it is time to turn on the Autostar and initialize. You may think you are "home free" at this point now that your ETX is mechanically perfect and you have sequentially and correctly downloaded the Autostar and the ETX has been placed in perfect Home Position. But you're not there yet.

You must still do the routine nightly initializing of the Autostar prior to accurate alignment of the telescope. There are some pitfalls in this, just as there have been elsewhere, and some things that should never be overlooked.

1. Turn on the power and you will see the copyright date and version of the Autostar come up.

2. Unless you have disabled it, the Sun warning appears; press 5 on the key pad to bypass this.

3. The Getting Started introduction comes up (unless disabled) and scrolls by; press ENTER to bypass this.

4. The last date you initialized and observed is displayed. Enter today's date.

5. Enter today's time in local 12-hour time (unless you have set the Autostar to use the 24-hour clock display). Set the time on Autostar about a minute ahead and use an accurate watch to signal when to press ENTER. Entering accurate time is very important. You can obtain time to within 0.7 second accuracy from the WWV Web Site (www.boulder.-nist.gov/timefreq/) or from the WWV short-wave radio station. It is very important that your time be within at least two minutes of accuracy for proper computer function.

6. On the same step, you must assure that you are observing in either AM or PM (default is PM; unless set to 24-hour); simply use the right and left arrow

keys to move to highlight and then scroll to your choice and press ENTER.

7. The Autostar then asks if you are in "Daylight Saving Time"; scroll and select appropriate response and press ENTER.

8. Then the Autostar display will show SETUP: Align. You are now ready to computer align your telescope.

So far, this is right out of the book. But wait a minute; slow down. You need to check a couple of things first. Among the most embarrassing and most difficult problems to diagnose are those that are the easiest to check. Let's make sure your settings are correct on the Autostar.

1. Check under SETUP/TELESCOPE:
   (a) Under MODEL, is your scope correctly selected (marked by a ">")? Display the correct model and press ENTER to select it.
   (b) Under MOUNT, do you have it set for whichever – polar or alt/az – that you plan to use? Again, scroll to the correct one and press ENTER.
   (c) Now scroll to TRACKING and check to assure that you are on SIDEREAL RATE.

2. Press MODE and scroll back to SETUP/SITE. Under SITE/SELECT press ENTER and see which of your observing sites has been selected. You would be surprised how easy it is to accidentally select the wrong site or to forget to change from the last site you observed from. This can make a clear difference in accuracy.

3. If you have time, you also might choose to double-check your Altitude and Azimuth Percent settings (see Part 2). I have found much success with the Azimuth Percent set to 15% and no change (01%) for Altitude.

A Quick Word About Computer Alignment

Of all the instructional data, Meade do a pretty good job of explaining proper access, centering, and selection of alignment stars for computer alignment in either observing mode; so that is not discussed here. There are some more enhancement points to ease some of your troubles, however. These are yet again some of those easy-to-overlook things that we forget when in a hurry.

1. Always use TWO STAR alignment, whether the EASY or USER-DEFINED. The two-star is just as simple, although it requires a bit more time, but is far more precise in the long run.

2. Make sure the axis clamps are secure and can firmly hold both axes while slewing. If they are loose, the Autostar "thinks" the telescope is covering more sky when it actually is slipping and standing still.

3. You may need to select alternate stars during your alignment only if:
   (a) the second star selected (if using EASY) is obscured by a terrestrial object;
   (b) the second star is located in the far north sky (or south, if in the southern hemisphere) near the celestial pole (which diminishes alignment accuracy);
   (c) the second star is far in the sky from the first.

4. To choose alternate stars (if using EASY), let the telescope access the star first but do not center and do not push ENTER if you do not want to use the star. After the beep merely tap your SCROLL key (lower right) to access the next recommended star. If you do not like that star again, keep scrolling until a suitable star far away from the first is found. Proceed as instructed.

5. Here are some points on alignment accuracy and using alternate stars when using High Precision and anytime you are using alternative stars. The more alternate stars that the Autostar must select for you, the less accurate is the GOTO accuracy. I do not have an explanation; but this has proved to be the case and is probably related to a time delay in the actual activation of the sidereal motor(s), since they do not commence until after Alignment Successful shows on the Autostar display.

6. If you are having obvious problems in alignment, for example, you see that the telescope has slewed to your first star and it is way off from the target star, you need to start over. Simply turn off the power, reinitialize (as described in the preceding pages), reset the scope to the Home Position and align anew. Sometimes the smallest error (again user error) can cause the greatest offsets in GOTO accuracy. If this is the case, it is best to reboot and start from scratch.

7. With the new versions and downloaders for the Autostar, there is a tendency during alignment (particularly after centering the second star and pressing ENTER) for the telescope to suddenly slew a very short and quick burst in RA (azimuth) after beeping. This is particularly true in polar mode. This is not a problem so do not recenter the alignment star. Merely commence to your first GOTO object. This slew appears to be (although I do not have confirmation of this) an adjustment that is automatically applied by the Autostar to compensate for the time you fiddled around getting the tracking motors engaged after the final alignment.

You have finally completed all the checklists. It seems like a lot of work doesn't it? Well, so did that desktop computer when you first started using it. It was as strange and unfamiliar as the day is long, but look at you now. The process is automatic; you log on without consciously thinking about it, you can manipulate files, create graphics, and surf the Internet without giving it a second thought. But it has not always been that way. Just as with your computer, so will it be with the ETX and the Autostar.

Once all of the procedures become implanted firmly within your mind, you will have an enhanced telescope that will slew, track, and target with the best of them. It takes a lot of work to get there; but the freedom from frustration, the joy of accessing thousands of objects the first time, the feeling of actually using your telescope instead of working with it (isn't that what jobs are for?) all combine to make this effort seem minor by comparison.

Good luck as you use these tips to attain the "Perfect Enhanced ETX Telescope."

Summary

I hope this book has shown just what you can do with the ETX. It may be small and have some limitations, but it is still a very capable telescope system for its price. Do not be afraid to try something with the ETX. If you proceed cautiously and use some common sense, you may just get more from it than you expected. Using my ETX-90RA in its "astrophotography configuration" as an example (Figure S.1), it really is the "Mighty ETX".

Figure S.1.
A "Mighty ETX" loaded with many accessories.

Appendix: Resources

Books

There are many excellent books on astronomy for the amateur. I will only mention a few here. Some of these books mention other books that are just as useful.

Astronomy with Small Telescopes, edited by Stephen F. Tonkin, has already been mentioned. It is a worthwhile addition to any small-telescope user's library. There are tips and techniques from real amateur astronomers who use telescopes ranging in size from 60 mm to 5 inches. Whether you use the equipment modifications or the deep sky observing techniques, there is something (and probably lots of "somethings") in this book that you will find valuable. As previously noted, the chapter on the ETX has been expanded into the book you hold in your hands.

Practical Astrophotography by Jeffrey R. Charles, another book in Springer's Practical Astronomy Series, has a lot of information and techniques that are applicable to photography with the ETX. If you plan to get serious about astrophotography with your ETX or any telescope, you will want this book.

AstroFAQs by Stephen F. Tonkin, also in Springer's Practical Astronomy Series, contains a lot of general information that answers many of the questions that new (and experienced) amateur astronomers will have at some point.

Turn Left at Orion: A Hundred Night Sky Objects to See in a Small Telescope – and How to Find Them by Guy Consolmagno, Dan M. Davis, Karen Kotash Sepp, and Ann Drogin. This book has been described as possibly the only book that many amateur astronomers will need. It covers the moon, with maps; the planets; objects that can be seen at each season throughout the year; use of a telescope; and more.

Norton's Star Atlas has been a standard reference for amateurs for decades. I purchased my first copy in the early 1960s when I was a teenager and used this copy frequently until 2000, when I (finally) purchased a more current edition. It contains star charts, tables of objects, definitions, and more. Highly recommended for any amateur astronomer.

The Trained Sky Star Atlas from Rigel Systems can be defined as a "streamlined Norton's". It contains star charts similar to those in Norton's and has tables showing bright stars (brighter than magnitude 2.5), brighter multiple stars, nebulae, galaxies, and star clusters. It is intended to really just help you navigate around the night sky. For anyone on a limited budget, this book, which is lower-priced than Norton's, may be just what you need.

Magazines

There are three monthly magazines that provide a lot of useful information (news, observing charts, product reviews, and more). *Sky & Telescope* from Sky Publishing Corporation (http://www.skypub.com) has been published the longest (since 1941; my subscription goes back to January 1960, and I proudly still have all these issues!).

Astronomy from Kalmbach Publishing Company (http://www.astronomy.com) is similar in purpose, although I find the "tone" of the magazine more oriented to popular astronomy, whereas *Sky & Telescope* covers all levels of astronomy and space science. Browse a couple of issues of both at your local library, newsstand, or bookstore, and see which suits you.

Astronomy Now (http://astronomynowstore.com) is Britain's leading astronomy magazine and has been published for over 10 years.

Another magazine that is totally oriented to the amateur astronomer is, not surprisingly, called *Amateur Astronomy* (http://www.amateurastronomy.com). This quarterly magazine has articles written by amateur astronomers for amateur astronomers. Some of the articles are highly technical or of limited use to the amateur with a small telescope such as the ETX. There are tidbits, technical or not, that can only be gleaned from this magazine.

Web Sites

With the growth in the World Wide Web over the last several years, Sites related to astronomy have grown in number. Many Sites are from established astronomical entities (universities, observatories, organizations, telescope manufacturers, and publishers) or from amateur astronomers. There is no complete collection of these Sites nor can there ever be with the dynamics of the Web. However, many Sites listed here have links to other Sites (that is why it is a "web"). Start surfing if you want to learn more about your telescope and astronomy. If you have a computer and access to the Web, you will find that a lot of information is available there.

One place to start is with my ETX Site (http://www.weasner. com/etx). The "Astronomy Links" page has links to many Sites on general astronomy, specific Sites on the ETX and other small telescopes, magazines, software, and dealers.

Another place to start is your telescope manufacturer's Web Site. For example, Meade Instruments (http://www.meade.com) or Celestron International (http://www.celestron.com) have information about their products, dealers, and more.

There is a Web Site that is geared to the Meade line of telescopes. Check MAPUG.com for the "Meade Advanced Products User Group". There is a lot of information, some technical, available through this Site, along with an archive of the MAPUG mailing list discussions.

The magazine Sites mentioned earlier have monthly observing information, tips, and other resources online.

Newsgroups

Newsgroups have been around for many years, longer than the Web. These provide near real-time discussion areas on a variety of topics. Probably the most popular one is the sci.astro.amateur newsgroup. This newsgroup has the most discussions, valuable responses from knowledgeable people to many of the questions raised, and a good "signal-to-noise" ratio. Another newsgroup that ETX users will want to monitor is alt.telescopes.meade. This one frequently has ETX-related content.

Related to newsgroups are "discussion groups" and "mailing lists". These come and go rather frequently, so check the links on my ETX Web Site for the latest groups and lists that discuss the ETX.

User Groups and Clubs

In many locations around the world, amateur astronomers frequently get together to discuss equipment and techniques. Ask your local dealers or check the "Resources" on the *Sky & Telescope* Web Site for a group or club near you. Besides monthly meetings there are also "star parties" and other observing events. These are usually discussed in the monthly magazines and on Web Sites.

Dealers

There are good, knowledgeable dealers and there are bad ones. Ask around, check the various newsgroups, and visit user Web Sites. Many excellent dealers have been conducting business for years, and some are recent additions. Some may be local to you or do all or most of their business through mail-order and online. Having a local dealer means you have someone you can talk to eyeball-to-eyeball and some place to easily make exchanges if something does not work out right for you. If you elect to use the good mail-order and online dealers, you should at least have an understanding of what you want to order. Return shipping charges can be an expense you would rather not deal with. Check the listing on the "Astronomy Links" page on my ETX Site for Web addresses of known reputable dealers.

Computer Software

Astronomy software for your desktop or laptop computer comes in four types: educational software, planetarium (star charting) software, telescope control software, and astrophotography software. One or more of these may be useful to you at some point.

Educational software teaches you something about astronomy. Being a long-term amateur astronomer, I have never had the need nor opportunity to use this type of software. So you are on your own here.

Planetarium software is available for Macintosh and Windows. Some popular commercially available examples are "Starry Night" from Space Software (http://www.starry night.com), "SkyMap Pro" from World Wide Software Publishing (http://www.wwsoftware.com), and "Voyager III" from Carina Software (http://www.CarinaSoft.com/). There are also some excellent freeware and shareware ones: "Hallo northern sky" by Han Kleijn (http://ourworld.compuserve. com/homepages/han_kleijn/software.htm), "Cartes du Ciel Sky Charts" by Patrick Chevalley (http://www.stargazing.net/ astropc/), and "Home Planet" by John Walker (http:// www.fourmilab.com/homeplanet/homeplanet.html). There is also software for Palm OS and WindowsCE handheld computers. Some examples are "Pocket Star Chart" from Muse of Fire (http://www.muse-of-fire.com/PSC.html), "Planetarium" by AHo Software (http://www.aho.ch/pilotplanets/), and "Star Pilot" by Star Pilot Technologies (http://www.star-pilot.com/). Some are free, and some are commercial products.

There is software that can control some telescopes. Since the Autostar for the ETX models is generally compatible with

Meade LX200 commands, many of the "plug-ins" that work with the LX200 will work with the ETX and Autostar. Check with the software developer to see whether they have a version that will work with your telescope.

Finally there is software designed for manipulation of CCD images and other astrophotographs. Probably the best program to "stack" multiple images into a single high-quality image is "AstroStack" by R.J. Stekelenburg (http://utopia.ision. nl/users/rjstek/). Other software typically comes with the CCD or imager you have purchased.

See the Astronomy Links page on my ETX Site for more links to astronomy software.

Biographies

Mike Weasner is the Webmaster of Weasner's Mighty ETX Site (http://www.weasner.com/etx). Mike grew up in Seymour, Indiana, USA. A brother started him in astronomy when he was six years old. Mike earned an astrophysics degree from Indiana University, Bloomington, Indiana; did some post-graduate work in meteorology at the University of Wisconsin, Madison; and served in the United States Air Force as a fighter pilot, an instructor, and a manager in the Air Force's Space Shuttle Program Office. He now works at an aerospace company. His hobbies include astronomy, science fiction, and computers. He enjoys classical music, science fiction movie soundtracks, and old radio shows such as *X Minus One* and *The Lone Ranger*. This book is Mike's second appearance as an author in the Practical Astronomy Series.

Contributors

Richard Seymour was born near Boston, Massachusetts, USA, and was fortunate in having parents who fostered an early interest in astronomy (hot chocolate dispensed during a 2am lunar eclipse, in a sleet storm. Thanks, Mom!). He still has his first 3-inch A.C. Gilbert $18 reflector and the Olcott and Mayall's *Field Guide to the Skies* he bought to go with it. An electrical engineer by training, a computer systems manager in a physics facility by trade, his odd bent towards delving into what makes his ETX-90's Autostar tick (an 8 MHz crystal, actually) has led him to try to help others via the medium of Mike's ETX Web Site.

P. Clay Sherrod is an author of many books about science, ranging from archeology and environmental and medical testing to astronomy. He has devoted his life and career to promoting the excitement and knowledge of astronomy and related sciences to students of all ages and interest. Sherrod has taught astronomy and related sciences at the university level for many years but lately is particularly focused on astronomical involvement with the lay public and community groups. Over his career Dr. Sherrod has presented thousands of lectures and popular syndicated writings on astronomy and archeology throughout North America to groups ranging from university scholars to elementary school and scouting groups. Now retired but still in demand for his services, Sherrod works closely with the astronomy community (public, professional, and educational) from a purely volunteer basis to provide the latest in astronomical equipment, research, and techniques through his private research, publications, and seminars/lectures nationwide. He maintains the Arkansas Sky Observatory, Conway/Petit Jean Mountain, established in 1971, and continues to provide modern astronomical research on planetary atmospheres, comet evolution, and cataclysmic variable stars.

Index